Role of DNA Methyltransferases in the Epigenome

Role of DNA Methyltransferases in the Epigenome

Special Issue Editors

Albert Jeltsch
Humaira Gowher

MDPI • Basel • Beijing • Wuhan • Barcelona • Belgrade

Special Issue Editors

Albert Jeltsch
Institute of Biochemistry and Technical Biochemistry
Department of Biochemistry
University of Stuttgart
Germany

Humaira Gowher
Department of Biochemistry
Purdue University
USA

Editorial Office
MDPI
St. Alban-Anlage 66
4052 Basel, Switzerland

This is a reprint of articles from the Special Issue published online in the open access journal *Genes* (ISSN 2073-4425) from 2018 to 2019 (available at: https://www.mdpi.com/journal/genes/special_issues/DNA_methyltransferases_Epigenome).

For citation purposes, cite each article independently as indicated on the article page online and as indicated below:

LastName, A.A.; LastName, B.B.; LastName, C.C. Article Title. *Journal Name* **Year**, *Article Number*, Page Range.

ISBN 978-3-03928-020-9 (Pbk)
ISBN 978-3-03928-021-6 (PDF)

Cover image courtesy of Albert Jeltsch.

© 2020 by the authors. Articles in this book are Open Access and distributed under the Creative Commons Attribution (CC BY) license, which allows users to download, copy and build upon published articles, as long as the author and publisher are properly credited, which ensures maximum dissemination and a wider impact of our publications.

The book as a whole is distributed by MDPI under the terms and conditions of the Creative Commons license CC BY-NC-ND.

Contents

About the Special Issue Editors vii

Albert Jeltsch and Humaira Gowher
Editorial—Role of DNA Methyltransferases in the Epigenome
Reprinted from: *Genes* **2019**, *10*, 574, doi:10.3390/genes10080574 1

Wendan Ren, Linfeng Gao and Jikui Song
Structural Basis of DNMT1 and DNMT3A-Mediated DNA Methylation
Reprinted from: *Genes* **2018**, *9*, 620, doi:10.3390/genes9120620 5

Albert Jeltsch, Julian Broche and Pavel Bashtrykov
Molecular Processes Connecting DNA Methylation Patterns with DNA Methyltransferases and Histone Modifications in Mammalian Genomes
Reprinted from: *Genes* **2018**, *9*, 566, doi:10.3390/genes9110566 25

Marthe Laisné, Nikhil Gupta, Olivier Kirsh, Sriharsa Pradhan and Pierre-Antoine Defossez
Mechanisms of DNA Methyltransferase Recruitment in Mammals
Reprinted from: *Genes* **2018**, *9*, 617, doi:10.3390/genes9120617 45

Si Xie and Chengmin Qian
The Growing Complexity of UHRF1-Mediated Maintenance DNA Methylation
Reprinted from: *Genes* **2018**, *9*, 600, doi:10.3390/genes9120600 63

Christian Bronner, Mahmoud Alhosin, Ali Hamiche and Marc Mousli
Coordinated Dialogue between UHRF1 and DNMT1 to Ensure Faithful Inheritance of Methylated DNA Patterns
Reprinted from: *Genes* **2019**, *10*, 65, doi:10.3390/genes10010065 75

Yang Zeng and Taiping Chen
DNA Methylation Reprogramming during Mammalian Development
Reprinted from: *Genes* **2019**, *10*, 257, doi:10.3390/genes10040257 89

Hemant Gujar, Daniel J. Weisenberger and Gangning Liang
The Roles of Human DNA Methyltransferases and Their Isoforms in Shaping the Epigenome
Reprinted from: *Genes* **2019**, *10*, 172, doi:10.3390/genes10020172 107

Allison B. Norvil, Debapriya Saha, Mohd Saleem Dar and Humaira Gowher
Effect of Disease-Associated Germline Mutations on Structure Function Relationship of DNA Methyltransferases
Reprinted from: *Genes* **2019**, *10*, 369, doi:10.3390/genes10050369 125

About the Special Issue Editors

Albert Jeltsch studied Biochemistry in Hannover, where he received his Ph.D. in 1994 working on restriction endonucleases. Following this, he started work on DNA methyltransferases and then moved to Giessen, where he received his Habilitation in 1999. He was appointed as Assistant Professor in Giessen, and as Associated Professor in Biochemistry at Jacobs University Bremen in 2003. He was appointed Full Professor in Biochemistry in 2006 before moving to University of Stuttgart in 2011, where he is currently Head of the Institute of Biochemistry and Technical Biochemistry.

The group of Prof. Jeltsch are leaders in studying the biochemistry and enzymology of DNA methyltransferases. They have long-standing expertise in the field of rational and evolutionary protein design of DNA-interacting enzymes and in the design of chimeric methylation enzymes for gene regulation in eukaryotic cells. In addition, their research covers molecular epigenetics, where they study the specificity and activity of histone methyltransferases and methyllysine reader domains, and have provided seminal discoveries in both fields. Prof. Jeltsch has published more than 250 scientific papers, many of which are in international leading journals. His work has amassed >17,500 citations, corresponding to an h-index of 67. He received the Gerhard Hess Award of the DFG in 1999 and the BioFuture Award of the Federal Minister of Research and Education (BMBF) in 2001. He is a member iof several editorial boards (including *Nucleic Acids Res.*, *Scientific Reports*, *ChemBioChem*, *Clinical Epigenetics*, *PLoS ONE*, *Genes*, and *Biochimie*) and is engaged in several university committees.

Humaira Gowher received her Master of Science degree in Biochemistry from Aligarh Muslim University, India, and Ph.D. from Justus Liebig University, Germany. After a postdoctoral appointment with Dr. Gary Felsenfeld at National Institutes of Health, Dr. Gowher joined the Department of Biochemistry at Purdue University, Indiana, USA, as Assistant Professor in the fall of 2013. Here, she was tenured and promoted to Associate Professor in 2019.

The overarching goal of her lab is to determine how epigenetic mechanisms, particularly DNA methylation, control cell identity and how these mechanisms are disrupted in various cancers. Her lab uses mouse embryonic stem cells and embryonal carcinoma cells as a model system to elucidate epigenetic mechanisms that control the activity of distal regulatory elements, enhancers, and insulators of developmental genes. The lab is currently funded by the National Institutes of Health, National Science Foundation, and American Heart Association. Dr Gowher is a member of the Purdue University Center for Cancer Research. She received the Showalter Trust Award from Purdue University in 2015, and Scientist Development Award from the American Heart Association in 2017. She serves as an ad hoc panelist for the NIH study sections Molecular Genetics B (MGB) and Development 2 (Dev2).

Editorial

Editorial—Role of DNA Methyltransferases in the Epigenome

Albert Jeltsch [1,*] and Humaira Gowher [2,*]

1. Department of Biochemistry, Institute of Biochemistry and Technical Biochemistry, University of Stuttgart, 70569 Stuttgart, Germany
2. Department of Biochemistry, Purdue University, West Lafayette, IN 47907, USA
* Correspondence: albert.jeltsch@ibtb.uni-stuttgart.de (A.J.); hgowher@purdue.edu (H.G.)

Received: 18 July 2019; Accepted: 25 July 2019; Published: 30 July 2019

Abstract: DNA methylation, a modification found in most species, regulates chromatin functions in conjunction with other epigenome modifications, such as histone post-translational modifications and non-coding RNAs. In mammals, DNA methylation has essential roles in development by orchestrating the generation and maintenance of the phenotypic diversity of human cell types. This Special Issue of Genes contains eight review articles, which cover several aspects of epigenome regulation by DNA methyltransferases (DNMTs), the enzymes responsible for the introduction of DNA methylation. The manuscripts present the most recent advances regarding the structure and function of DNMTs, their targeting and regulation by interacting factors and chromatin modifications, and the roles of DNMTs in mammalian development and human diseases. However, many aspects of these important enzymes are still insufficiently understood. Potential directions of future work are the regulation of DNMTs by post-translational modifications and their connection to cellular signaling and second messenger cascades on one hand and to large multifactorial epigenetic chromatin circuits on the other. Additionally, technical advancements, including the availability of designer nucleosomes and the rapid development of cryo-electron microscopy are expected to trigger breakthrough discoveries in this exciting field.

Keywords: DNA methyltransferase function; DNA methyltransferase mechanism; DNA methyltransferase regulation; DNA methyltransferase structure; DNMT1; DNMT3A; DNMT3B; DNA Methylation

1. Introduction

DNA methylation at the cytosine—C5 position is found in many species. The methylation is placed in the major groove of double-stranded B-DNA, where it does not interfere with the Watson/Crick base pairing, but it can influence the binding of proteins to specific DNA sequences and thereby, for example, direct the binding of transcription factors to gene regulatory elements. By this mechanism, the methylation adds extra information to the DNA that is not encoded in the DNA sequence and represents one important component of the epigenome [1]. DNA methylation regulates several chromatin functions in conjunction with other epigenome modifications, such as histone post-translational modifications and non-coding RNAs [2]. By orchestrating the generation and maintenance of the phenotypic diversity of the various cell types of the body, DNA methylation plays an essential role in mammalian development [3]. Moreover, DNA methylation provides the substrate for more recently discovered Ten-eleven Translocation (TET) enzymes, which oxidize 5-methylcytosine to the hydroxyl, formyl, and carboxyl state [4]. Numerous studies have demonstrated that aberrant DNA methylation has serious consequences, including the onset and progression of cancer [5,6]. DNA methyltransferases, the enzymes that introduce DNA methylation, clearly are one of the key players in molecular epigenetics [7]. Despite being studied for more than 40 years [8–10], recent work has brought important advances in our understanding of the mechanism, function, and regulation of DNA methyltransferases some of which are collected and reviewed in eight publications in this special issue of Genes.

2. Structure and Function of DNMTs

Ren et al. [11] describe the newest discoveries regarding the structural basis of DNA methyltransferase 1 (DNMT1) and DNMT3A mediated DNA methylation. Based on recent structure-function investigations of the individual domains or large fragments of DNMT1 and DNMT3A, they review the molecular basis for their substrate recognition and specificity, intramolecular domain–domain interactions, as well as their crosstalk with other epigenetic mechanisms. Their paper highlights the multifaceted nature of the regulation of both DNMT1 and DNMT3A/3B, which is essential for the precise establishment and maintenance of lineage-specific DNA methylation patterns.

3. Chromatin Recruitment and Regulation of DNMTs

Jeltsch et al. [12] describe the genomic distribution and variability of DNA methylation in different genomic elements in human and mouse DNA and the connection of DNA methylation with several key histone post-translational modifications, including methylation of H3K4, H3K9, H3K27, and H3K36, and also with nucleosome remodeling. Based on this, they review the mechanistic features of mammalian DNA methyltransferases and their associated factors that recruit these enzymes to genomic sites and mediate the crosstalk between DNA methylation and chromatin modifications.

Laisne et al. [13] describe our current understanding of the recruitment mechanisms of DNA methyltransferase to target sites in mammals. This includes mechanisms of DNMT recruitment by transcription factors, other interacting chromatin modifiers and by RNA. These mechanisms are presented in the context of biologically relevant epigenetic events illustrating how the specific recruitment of DNMTs controls epigenetic signaling.

Xie and Qian [14] and Bronner et al. [15] focus on the specific question of the complex role of UHRF1 in the regulation and targeting of DNMT1. UHRF1 has previously been reported to regulate DNMT1 in multiple ways, including control of substrate specificity and activity based on allosteric regulation of DNMT1, as well as histone and DNMT1 ubiquitylation. Moreover, UHRF1 contributes to the proper genome targeting of DNMT1 by several chromatin interactions with hemimethlyated DNA and modified histone tails. The interplay of these complex multidomain proteins is one illustrative example of the complexity of epigenetic regulation cascades.

4. Role of DNMT in Development and Disease

Zeng and Chen [16] integrate these views and review the process of DNA methylation reprogramming during mammalian development. They describe the two waves of DNA methylation reprogramming in mammals occurring in the germline and after fertilization and explain their mechanistic underpinnings. By this they provide an overview of these key reprogramming events, focusing on the important players in these processes including DNA methyltransferases (DNMTs) and TET family of 5mC dioxygenases. Gujar et al. [17] review the role of isoforms of human DNMTs in shaping the epigenome mainly focusing on the DNMT3B isoforms which have documented roles in development and cancer progression, and add another layer of complexity to epigenetic regulation in biological systems.

Norvil et al. [18] describe the effect of disease-associated germline mutations in DNMTs. Recent advances in whole genome association studies have helped to identify mutations and genetic alterations of DNMTs in various diseases that have the potential to affect the biological function and activity of these enzymes. Several of these mutations are germline-transmitted and associated with a number of hereditary disorders, including neurological dysfunction, growth defects, and inherited cancers. This review describes DNMT mutations that are associated with rare diseases, the effects of these mutations

5. Conclusions and Outlook

While this collection of review articles illustrates in an impressive manner the high level of mechanistic understanding of DNMTs that has been reached over the last decade of research, it also emphasizes the gaps and open questions in the field that need to be answered. It is anticipated that, in future, the investigation of the targeting and regulation of DNMTs will be intensified to finally understand the mechanisms leading to the generation of DNA methylation patterns during early development, germ cell development and onset of disease. In this respect, the details of the regulation of DNMTs by post-translational modifications are still uncovered, as well as their connection to cellular signaling and second messenger cascades. The increasing availability of designer nucleosomes will allow powerful enzymatic in vitro studies regarding the recruitment and crosstalk of DNMTs with other chromatin marks. The further improvement of cryo-electron microscopy will enable structural investigation of larger protein complexes with atomic resolution, allowing to study the structure and conformation of DNMTs also in complex with their regulatory factors.

Funding: Work in the Gowher lab is supported by NIHR01GM118654-01 and NSF 1,716,678 grants. Related work in the Jeltsch lab is supported by the DFG (JE 252/6 and JE 252/15).

Conflicts of Interest: The authors declare no conflict of interest.

References

1. Allis, C.D.; Jenuwein, T. The molecular hallmarks of epigenetic control. *Nat. Rev. Genet.* **2016**, *17*, 487–500. [CrossRef] [PubMed]
2. Schübeler, D. Function and information content of DNA methylation. *Nature* **2015**, *517*, 321–326. [CrossRef] [PubMed]
3. Feng, S.; Jacobsen, S.E.; Reik, W. Epigenetic reprogramming in plant and animal development. *Science* **2010**, *330*, 622–627. [CrossRef] [PubMed]
4. Wu, X.; Zhang, Y. TET-mediated active DNA demethylation: Mechanism, function and beyond. *Nat. Rev. Genet.* **2017**, *18*, 517–534. [CrossRef] [PubMed]
5. Baylin, S.B.; Jones, P.A. A decade of exploring the cancer epigenome—Biological and translational implications. *Nat. Rev. Cancer* **2011**, *11*, 726–734. [CrossRef] [PubMed]
6. Feinberg, A.P.; Koldobskiy, M.A.; Göndör, A. Epigenetic modulators, modifiers and mediators in cancer aetiology and progression. *Nat. Rev. Genet.* **2016**, *17*, 284–299. [CrossRef] [PubMed]
7. Jeltsch, A.; Jurkowska, R.Z. *DNA Methyltransferases—Role and Function*; Springer-Nature: Berlin/Heidelberg, Germany, 2016.
8. Jeltsch, A. Beyond Watson and Crick: DNA Methylation and Molecular Enzymology of DNA Methyltransferases. *ChemBioChem* **2002**, *3*, 274–293. [CrossRef]
9. Jurkowska, R.Z.; Jurkowski, T.P.; Jeltsch, A. Structure and Function of Mammalian DNA Methyltransferases. *Chembiochem* **2011**, *12*, 206–222. [CrossRef] [PubMed]
10. Jurkowska, R.Z.; Jeltsch, A. Mechanisms and Biological Roles of DNA Methyltransferases and DNA Methylation: From Past Achievements to Future Challenges. *Results Probl. Cell Differ.* **2016**, *945*, 1–17.
11. Ren, W.; Gao, L.; Song, J. Structural Basis of DNMT1 and DNMT3A-Mediated DNA Methylation. *Genes* **2018**, *9*, 620. [CrossRef] [PubMed]
12. Jeltsch, A.; Broche, J.; Bashtrykov, P. Molecular Processes Connecting DNA Methylation Patterns with DNA Methyltransferases and Histone Modifications in Mammalian Genomes. *Genes* **2018**, *9*, 566. [CrossRef] [PubMed]
13. Laisné, M.; Gupta, N.; Kirsh, O.; Pradhan, S.; Defossez, P.A. Mechanisms of DNA Methyltransferase Recruitment in Mammals. *Genes* **2018**, *9*, 617. [CrossRef] [PubMed]
14. Xie, S.; Qian, C. The Growing Complexity of UHRF1-Mediated Maintenance DNA Methylation. *Genes* **2018**, *9*, 600. [CrossRef] [PubMed]

15. Bronner, C.; Alhosin, M.; Hamiche, A.; Mousli, M. Coordinated Dialogue between UHRF1 and DNMT1 to Ensure Faithful Inheritance of Methylated DNA Patterns. *Genes* **2019**, *10*, 65. [CrossRef] [PubMed]
16. Zeng, Y.; Chen, T. DNA Methylation Reprogramming during Mammalian Development. *Genes* **2019**, *10*, 257. [CrossRef] [PubMed]
17. Gujar, H.; Weisenberger, D.J.; Liang, G. The Roles of Human DNA Methyltransferases and Their Isoforms in Shaping the Epigenome. *Genes* **2019**, *10*, 172. [CrossRef] [PubMed]
18. Norvil, A.B.; Saha, D.; Dar, M.S.; Gowher, H. Effect of Disease-Associated Germline Mutations on Structure Function Relationship of DNA Methyltransferases. *Genes* **2019**, *10*, 369. [CrossRef] [PubMed]

© 2019 by the authors. Licensee MDPI, Basel, Switzerland. This article is an open access article distributed under the terms and conditions of the Creative Commons Attribution (CC BY) license (http://creativecommons.org/licenses/by/4.0/).

Review

Structural Basis of DNMT1 and DNMT3A-Mediated DNA Methylation

Wendan Ren [1,†], Linfeng Gao [2,†] and Jikui Song [1,2,*]

1. Department of Biochemistry, University of California, Riverside, CA 92521, USA; wendan@ucr.edu
2. Environmental Toxicology Program, University of California, Riverside, CA 92521, USA; lgao010@ucr.edu
* Correspondence: jikui.song@ucr.edu; Tel.: +1-951-827-4221
† These authors contributed equally to this work.

Received: 7 November 2018; Accepted: 4 December 2018; Published: 11 December 2018

Abstract: DNA methylation, one of the major epigenetic mechanisms, plays critical roles in regulating gene expression, genomic stability and cell lineage commitment. The establishment and maintenance of DNA methylation in mammals is achieved by two groups of DNA methyltransferases (DNMTs): DNMT3A and DNMT3B, which are responsible for installing DNA methylation patterns during gametogenesis and early embryogenesis, and DNMT1, which is essential for propagating DNA methylation patterns during replication. Both groups of DNMTs are multi-domain proteins, containing a large N-terminal regulatory region in addition to the C-terminal methyltransferase domain. Recent structure-function investigations of the individual domains or large fragments of DNMT1 and DNMT3A have revealed the molecular basis for their substrate recognition and specificity, intramolecular domain-domain interactions, as well as their crosstalk with other epigenetic mechanisms. These studies highlight a multifaceted regulation for both DNMT1 and DNMT3A/3B, which is essential for the precise establishment and maintenance of lineage-specific DNA methylation patterns in cells. This review summarizes current understanding of the structure and mechanism of DNMT1 and DNMT3A-mediated DNA methylation, with emphasis on the functional cooperation between the methyltransferase and regulatory domains.

Keywords: DNMT1; DNMT3A; DNA methyltransferase; maintenance DNA methylation; de novo DNA methylation; allosteric regulation; autoinhibition

1. Introduction

DNA methylation represents one of the major epigenetic mechanisms that critically influence gene expression and cell fate commitment [1–6]. In mammals, DNA methylation is essential for the silencing of retrotransposons [7–9], genomic imprinting [10,11] and X-chromosome inactivation [12,13]. Mammalian DNA methylation predominantly occurs at the C-5 position of cytosine within the CpG dinucleotide context, accounting for ~70–80% of CpG sites throughout the genome [14]. The establishment of DNA methylation is achieved by the closely related DNA methyltransferases 3A (DNMT3A) and 3B (DNMT3B), designated as de novo DNA methyltransferases, during germ cell development and early embryogenesis [15,16]. Subsequently, clonal transmission of specific DNA methylation patterns is mainly mediated by DNA methyltransferase 1 (DNMT1), designated as maintenance DNA methyltransferase, in a replication-dependent manner [17,18]. However, the classification of DNMT3A/3B as de novo methyltransferases and DNMT1 as maintenance DNA methyltransferase appears to be an oversimplification, as increasing evidence has revealed an important role of DNMT3A and DNMT3B in DNA methylation maintenance [19,20], while other studies have pointed to the de novo methylation activity of DNMT1 in specific loci [21,22]. A detailed understanding of the structure and regulation of DNMT1 and DNMT3A/3B is essential for elucidating their roles in DNA methylation maintenance and establishment in cells.

Both DNMT1 and DNMT3A/3B belong to the class I methyltransferase family [23], featured by a conserved catalytic core termed Rossmann fold, which consists of a mixed seven-stranded β-sheet flanked by three α-helices on either side [24]. These enzymes catalyze the methylation reaction in an S-adenosyl-L-methionine (AdoMet)-dependent manner, with the catalytic core harboring essential motifs for enzymatic catalysis and cofactor binding. In addition, a subdomain, termed target recognition domain (TRD), is inserted between the central β-sheet and the last α-helix of the catalytic core [24]. The TRD bears no sequence similarity between DNMT1 and DNMT3s; instead, it participates in DNA binding to ensure substrate specificity of each enzyme.

To ensure proper programming of DNA methylation patterns in cell linage commitment, the functions of DNMTs are subject to a stringent regulation during development [25,26]. Unlike their bacterial counterparts that contain only the methyltransferase (MTase) domain, both DNMT1 and DNMT3s are multi-domain proteins, containing a large regulatory region in addition to the C-terminal MTase domain (Figure 1) [18,27]. Recent studies have generated a large body of structural and functional information on both groups of enzymes, including the molecular basis underlying their enzyme-substrate recognition, and the regulatory roles of their N-terminal segments in the substrate specificity, enzymatic activity as well as genomic targeting. This review provides an overview on the recent progress in structural and mechanistic understanding of DNMT1 and DNMT3A, with an emphasis on how the regulatory and MTase domains of each enzyme cooperate in maintenance and de novo DNA methylation, respectively.

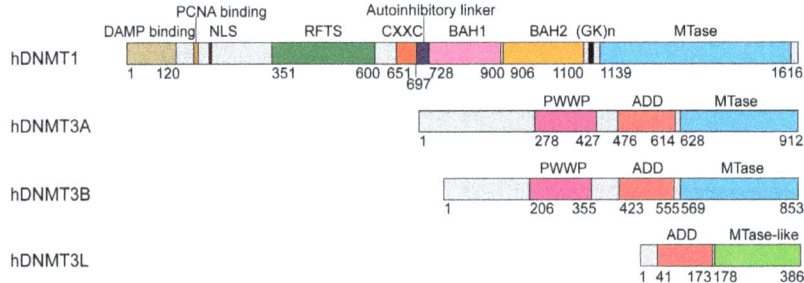

Figure 1. Domain architectures of human DNA methyltransferases: DNMT1, DNMT3A and DNMT3B, and regulator DNMT3L, with individual domains marked by residue numbers.

2. Structure and Mechanism of DNMT1

DNMT1 is comprised of ~1600 amino acids, with an N-terminal regulatory region covering two thirds of the sequence, a highly conserved (GK)n repeat and a C-terminal MTase domain (Figure 1). The regulatory region starts with a ~300 amino acid-long N-terminal domain (NTD) harboring a variety of protein and/or DNA interaction sites, followed by a replication foci-targeting sequence (RFTS) domain, a CXXC zinc finger domain and a pair of bromo-adjacent-homology (BAH) domains (Figure 1). The function of DNMT1 in replication-dependent DNA methylation maintenance is supported by its localization in replication foci during the S phase, and in vitro a 3–40 fold enzymatic preference for hemimethylated CpG sites [18,28], an epigenetic mark enriched at the replication foci [29]. How the regulatory domains of DNMT1 are coordinated in attaining its enzymatic and spatiotemporal regulations remains a long-lasting topic of interest. Nevertheless, recent structure-function studies of various DNMT1 fragments under different DNA binding states [30–33] have started to illuminate how different domains of this enzyme orchestrate its activity in maintenance DNA methylation.

2.1. Enzyme-Substrate Interaction of DNMT1

The crystal structure of a mouse DNMT1 fragment (mDNMT1, residues 731–1602) covalently bound to a 12-mer hemimethylated DNA duplex provides insight into the productive state of DNMT1

(Figure 2A) [31]. The DNA molecule contains one central CpG site in which a 5-methylcytosine (5mC) and a 5-fluorocytosine (5fC) were installed on the template and target strands, respectively (Figure 2B). The use of 5fC permits the formation of an irreversible, covalent complex between mDNMT1 and DNA [34]. The mDNMT1 fragment contains the pair of BAH domains (BAH1, BAH2) and the MTase domain.

The structure of the mDNMT1-DNA covalent complex reveals that the MTase domain, composed of a catalytic core and a large TRD (~200 amino acids), is organized into a two-lobe architecture, creating a cleft to harbor the DNA duplex (Figure 2A). The two BAH domains are separated by one α-helix, both with a tilted β-barrel fold that is reminiscent of other BAH domains (Figure 2A) [35]. Both BAH domains are structurally associated with the MTase domain, forming an integrated structural unit. The BAH1 domain is attached to the MTase domain through antiparallel β-pairing, as well as hydrophobic clustering, while the BAH2 domain interacts with the MTase domain mainly through hydrophobic contacts, with a long loop (BAH2-loop) protruding from one end of the β-barrel to join with the TRD at the tip (Figure 2A). This mDNMT1 construct also contains two Cys3His-coordinated zinc finger clusters, one located in the TRD while the other associates BAH1 with the subsequent α-helix (Figure 2A). The mDNMT1-DNA interaction spans eight base pairs, resulting in a buried surface area of ~2100 Å^2. The target cytosine, 5fC, is flipped out of the DNA duplex and inserts into the active site of mDNMT1, where it forms a covalent linkage with the catalytic cysteine C1229, leading to hydrogen bonding interactions with a number of highly conserved residues (Figure 2C). The base flipping of 5fC creates a large cavity at the hemimethylated CpG site, which is in turn filled with bulky side chains of K1537 from the TRD and W1512 from the catalytic core (Figure 2B). This protein-DNA intercalation further shifts the orphan guanine, which is otherwise paired with the flipped-out 5fC, one base down, resulting in the flipping out of a second nucleotide from the template strand (Figure 2B). The interaction of mDNMT1 with the hemimethylated CpG site involves two loops from the TRD (TRD loop I: Residues 1501–1516 and TRD loop II: Residues 1530–1537) and one loop from the catalytic site (catalytic loop: Residues 1227–1243). Toward the DNA major groove, residues from TRD loop I form a concave hydrophobic surface to harbor the methyl group of 5mC (Figure 2D). On the other hand, residues from TRD loop II engage in base-specific hydrogen bonding interactions with the CpG site (Figure 2E). On the minor groove side, residues from the catalytic loop also form base-specific contacts with the CpG site through hydrogen bonding interactions (Figure 2E). In addition, residues from both the TRD and catalytic core are involved in salt-bridge or hydrogen-bonding interactions with the DNA backbone. The two BAH domains are positioned distant to the DNA binding site. Nevertheless, residues from the tip of the BAH2-loop contribute to the DNA binding through hydrogen bonding interactions with the DNA backbone of the target strand (Figure 2A).

In summary, the structure of the productive mDNMT1-DNA complex provides the molecular basis for the substrate recognition of DNMT1. The extensive protein-DNA contacts underlie the processive methylation kinetics of this enzyme [36,37]. More importantly, it offers explanations on the strict substrate specificity of DNMT1 on the CpG sites, as well as on the marked substrate preference of DNMT1 toward hemimethylated CpG sites [18,28].

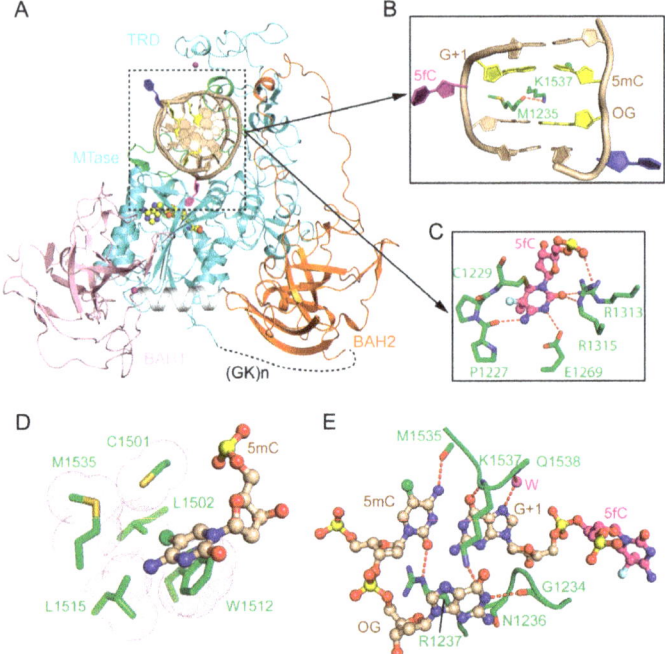

Figure 2. Structure of mDNMT1-DNA productive complex. (**A**) Structural overview of mDNMT1 (amino acids 731–1602) covalently bound to hemimethylated DNA (Protein Data Base (PDB) 4DA4). The zinc ions are shown in purple spheres. 5fC and another flipped-out cytosine from the template strand are colored in purple and blue, respectively. (**B**) The DNA cavity vacated by the base flipping is filled with mDNMT1 residues M1235 and K1537. (**C**) The flipped-out 5fC is surrounded by active site residues through covalent linkage or hydrogen bonding interactions. (**D**) Residues from the target recognition domain (TRD) loop II form a hydrophobic groove harboring the methyl group from 5mC. (**E**) CpG-specific interactions by the TRD loop I and the catalytic loop. 5mC: 5-methylcytosine; 5fC: 5-fluorocytosine.

2.2. CXXC Domain-Mediated Autoinhibition of DNMT1

The CXXC domain of DNMT1 belongs to one family of zinc finger domains that specifically bind to unmethylated CpG-containing DNA [30,38]. It manifests in a crescent-like fold, with two zinc finger clusters formed by the conserved CXXCXXC motifs in cooperation with distal cysteines. The crystal structure of an mDNMT1 fragment (residues 650–1602), spanning from the CXXC domain to the MTase domain, in complex with a 19-mer DNA duplex containing unmethylated CpG sites provides insight into the functional role of this domain (Figure 3A) [30]. In the structure, the CXXC domain is positioned on the opposite side of the MTase domain from the BAH domains, with a long CXXC-BAH1 domain linker (also known as autoinhibitory linker) running across the catalytic cleft (Figure 3A). The mDNMT1-unmethylated DNA complex contains two separate DNA-binding interfaces, one located in the CXXC domain and the other located in the MTase domain. At one end of the DNA, the CXXC domain interacts with the DNA molecule from both the major groove and the minor groove, with a loop segment (R684-S685-K686-Q687) penetrating into the CpG site for base-specific contacts (Figure 3B,C). At the other end of the DNA, the MTase domain interacts with the DNA backbone through the C-terminal portion of the catalytic loop (residues M1235, R1237 and R1241) and the adjacent α-helix (R1278 and R1279) (Figure 3D). These protein-DNA interactions together localize the DNA molecule outside the catalytic cleft, resulting in an autoinhibitory conformation of

DNMT1. Structural comparison of the autoinhibitory and active states of mDNMT1 reveals that the largest conformational change of mDNMT1 lies in the catalytic loop, which is poised in a retracted conformation in the autoinhibitory state, but penetrates into the DNA minor groove in the active state (Figure 3E). Furthermore, the α-helix following the catalytic loop undergoes a kinked-to-straight conformational transition, thereby regulating the contact between the catalytic loop and the DNA minor groove (Figure 3E). Indeed, a subsequent study indicated that disruption of this conformational transition leads to the impaired enzymatic activity of DNMT1 [39], highlighting the importance of this conformational switch in DNMT1-mediated DNA methylation.

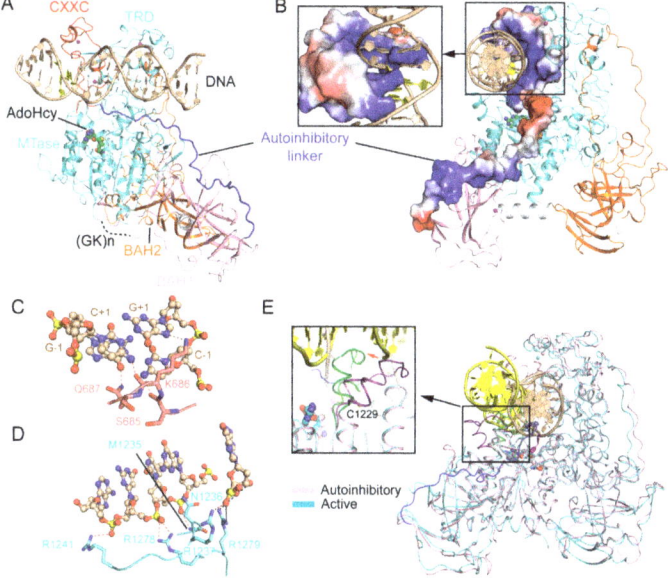

Figure 3. Structural analysis of the CXXC domain-mediated DNMT1 autoinhibition. (**A**) Structural overview of mDNMT1 (amino acids 650–1602) bound to a 19-mer DNA duplex containing unmethylated CpG sites (PDB 3PT6). (**B**) Surface views of the CXXC domain and the autoinhibitory linker in the complex of mDNMT1 with unmethylated CpG DNA. (**C**) Base-specific interactions between the CXXC domain and the CpG site. The hydrogen bonding interactions are depicted as dashed lines. (**D**) The MTase-DNA interactions in the autoinhibitory complex. (**E**) Structural overlay between the active (light blue) (PDB 4DA4) and autoinhibitory (pink) (PDB 3PT6) complexes of mDNMT1, with the catalytic loops highlighted in the expanded view.

These structural observations therefore led to an autoinhibitory model of DNMT1: The CXXC domain specifically interacts with the unmethylated CpG site, which in turn stabilizes the positioning of the autoinhibitory linker over the catalytic cleft, leading to the extrusion of the unmethylated CpG DNA from the catalytic site. This model therefore assigns a regulatory role to the CXXC domain in inhibiting the de novo methylation activity of DNMT1. Indeed, enzymatic assays based on the mDNMT1(650–1602) construct indicated that disruption of the CXXC-CpG interaction or deletion of the autoinhibitory linker both led to enhanced enzymatic activity of DNMT1 on unmethylated CpG DNA, but resulted in no significant change to hemimethylated substrates, lending support to the autoinhibitory mechanism. However, it is worth noting that a later study on full-length DNMT1 failed to identify any significant impact of the CXXC-DNA interaction on the substrate specificity of DNMT1 in vitro [40], suggesting that additional factors (e.g., protein interactions or post-translational modifications) may be needed to stabilize the CXXC domain-mediated autoinhibitory conformation, thereby ensuring the substrate specificity of DNMT1 in cells.

2.3. RFTS Domain-Mediated Autoinhibition of DNMT1

The crystal structures of DNA-free mouse and human DNMT1 fragments, spanning from the RFTS domain toward the MTase domain, reveal that the RFTS domain closely associates with the MTase domain, resulting in a compact fold (Figure 4A) [32,33]. In both structures, the RFTS domain folds into two lobes, separated by a 24-amino acid long α-helix (Figure 4A). The N-lobe is dominated by a zinc finger cluster, followed by a six-stranded β-barrel, while the C-lobe is assembled into a helical bundle (Figure 4A). The N and C lobes form an acidic cleft, where the linker sequence downstream of the RFTS domain extends away from the RFTS domain (Figure 4A). The intramolecular contact between the RFTS and MTase domains is underpinned by hydrogen bonding interactions between the residues from the C-lobe of the RFTS and the residues from the TRD (Figure 4B), which partially overlap with the DNA binding surface of the TRD (Figure 2A). The CXXC domain is positioned adjacent to the RFTS domain, adopting a conformation similar to its DNA-bound state (Figure 4A). Structural comparison of DNA-free DNMT1 and its unmethylated CpG DNA-bound state reveals a large conformational repositioning of the CXXC domain: It sits on one side of the TRD in the structure of mDNMT1–19-mer unmethylated CpG DNA, but moves to the front of the TRD in the structure of free DNMT1, resulting in a translocation of ~30 Å (Figure 4C). As a result, the autoinhibitory linker downstream of the CXXC domain undergoes a large conformational change between the two complexes: It runs across the catalytic cleft in the DNMT1-unmethylated CpG DNA complex but is released from the catalytic cleft in free DNMT1 (Figure 4C). Intriguingly, this repositioning of the autoinhibitory linker is accompanied by a loop-to-helix conformational transition: The N-terminal end of the linker assumes an extended conformation in unmethylated CpG-bound DNMT1 but shows a helical structure in free DNMT1 (Figure 4C). At the C-terminal end of this helix, residues D700 and E703 form salt bridges with residues R582 and K586 from the RFTS domain, while residue D702 forms hydrogen bonds with residues M1232 and N1233 from the catalytic core, which together help to strengthen the interaction between the RFTS and MTase domains (Figure 4D). Consistently, deletion of residues 701–711 from the autoinhibitory linker led to significantly enhanced enzymatic activities of DNMT1 [33]. These data therefore suggest that the autoinhibitory linker not only plays a critical role in the CXXC domain-mediated DNMT1 autoinhibition, but also contributes to the RFTS domain-mediated DNMT1 autoinhibition.

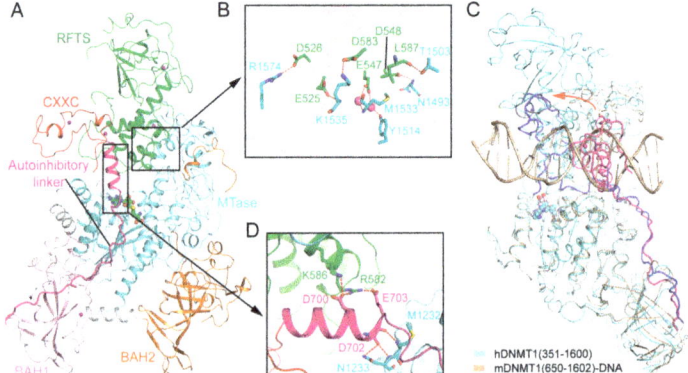

Figure 4. Structural analysis of the replication foci-targeting sequence (RFTS) domain-mediated DNMT1 autoinhibition. (**A**) Structural overview of hDNMT1 (amino acids 351–1602) (PDB 4WXX). (**B**) The intramolecular interactions between the RFTS (green) and MTase (aquamarine) domains. The hydrogen bonding interactions are depicted as dashed lines. The water molecules are shown as purple spheres. (**C**) Structural overlap between the CXXC (PDB 3PT6) and RFTS (PDB 4WXX) mediated autoinhibitory complexes, with the autoinhibitory linkers colored in blue and light magenta, respectively. The repositioning of the CXXC domain is indicated by a red arrow. (**D**) The interaction of the autoinhibitory linker (magenta) with both the RFTS (green) and MTase domains (aquamarine).

2.4. Allosteric Regulation of DNMT1

Crystal structures of DNMT1 in a DNA-free state, in complex with unmethylated CpG DNA and in complex with hemimethylated CpG DNA together demonstrate that DNMT1 may adopt distinct conformational states under different DNA binding conditions, suggesting a multi-layered regulation of DNMT1 activity. It is conceivable that the interconversion between these states permits DNMT1 to discriminate the DNA substrates under different epigenetic environments, such as methylation-free CpG islands compared to heavily methylated heterochromatic regions (Figure 5). The stabilization of each conformation is likely to be achieved by the distinct DNA or histone-binding mode of DNMT1 under different environments, ensuring DNMT1 will replicate the DNA methylation pattern both faithfully and efficiently. Indeed, emerging studies have suggested a model in which DNMT1 mediates region-specific DNA methylation maintenance, rather than site-specific DNA methylation maintenance [41].

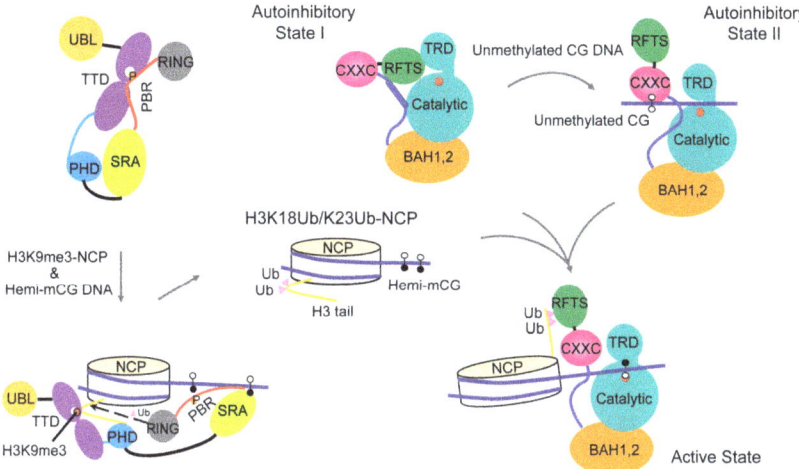

Figure 5. A model for the allosteric regulation of DNMT1-mediated maintenance DNA methylation. Hemimethylated DNA and histone H3K9me3 serve as epigenetic signals to promote UHRF1-mediated ubiquitination of histone H3, which in turn shifts the conformation of DNMT1 from the autoinhibitory state into an active state for maintenance DNA methylation. UHRF1 P656, which occupies the H3K9me3-binding cage of the tandem Tudor domain (TTD) in the closed UHRF1 conformation, is indicated by the letter P. The active site of DNMT1 is marked by a filled red circle.

The RFTS domain mediates the localization of DNMT1 to replication foci and constitutive heterochromatin from late S throughout the G2 and M phases [42,43]. A number of mutations in the RFTS domain have been associated with neurological disorders, including hereditary sensory autonomic neuropathy with dementia and hearing loss (HSAN1E) [44,45], cerebella ataxia, deafness and narcolepsy (ADCA-DN) [46,47]. These mutations presumably affect the folding and stability of the RFTS domain [33], which in turn may lead to the dysregulation of DNMT1-mediated methylation. Recent structural and functional characterizations of the interaction between the DNMT1 RFTS domain and histone modifications have further elucidated the functional implication of the RFTS domain-mediated DNMT1 autoinhibition [48]. In particular, it has been shown that the DNMT1 RFTS domain binds to histone H3 ubiquitinated at lysine 14 (K14Ub), 18 (K18Ub) and/or 23 (K23Ub), with a preference for H3 with two mono-ubiquitination (H3Ub2) [48–50]. The crystal structure of the RFTS domain of hDNMT1 in complex with H3-K18Ub/K23Ub reveals that the two ubiquitin moieties engage in hydrophobic interactions with two discrete surfaces of the N-lobe of RFTS, separated by a loop segment [48]. The N-terminal tail of H3 lies between the C-lobe and the ubiquitin molecule conjugated to H3K23, leading to the eviction of the linker sequence downstream of the RFTS domain out of the cleft between the N and C lobes [48]. In accordance with these structural changes, the interaction of DNMT1 RFTS with H3Ub2 results in a substantially elevated level of activity of DNMT1 [48], suggesting that H3Ub2 may serve as an epigenetic signal that relieves the RFTS-mediated autoinhibition of DNMT1. These studies have therefore established a link between the chromatin targeting and enzymatic activation of DNMT1, unveiling the molecular mechanism for RFTS regulation (Figure 5). It is worth noting that the H3 K14Ub/K18Ub/K23Ub marks are the enzymatic products of UHRF1 (ubiquitin-like, containing plant homeodomain (PHD) and RING finger domains) [48–50], a key regulatory protein of DNMT1-mediated maintenance DNA methylation [51,52]. UHRF1 is also a multi-domain protein comprised of an N-terminal ubiquitin-like (UBL) domain, a tandem Tudor domain (TTD), a plant homeodomain (PHD), a SET and RING-associated (SRA) domain and a C-terminal RING finger domain [53]. An intramolecular interaction between the TTD domain and the C-terminal polybasic region (PBR) of UHRF1 results in a closed conformation that occludes UHRF1 from chromatin association [54–57]. During the S phase, the association of UHRF1 with histone H3 trimethylated at lysine 9 (H3K9me3) [58–64], a silencing histone mark [65], and hemimethylated CpG DNA [51,52,61,66–69] leads to the conformational opening [54–56], and enhanced E3 ubiquitin ligase activity of UHRF1 (Figure 5) [70]. In this context, the DNMT1 RFTS domain serves as an effector module that transmits the H3K9me3 signal into DNMT1-mediated DNA methylation (Figure 5).

2.5. Regulatory Role of DNMT1 N-Terminal Domain

The N-terminal domain (NTD) appears not to affect the enzymatic activity of DNMT1. Instead, this region serves as a platform for the interaction between DNMT1 and proteins or DNA. Of particular note, the fragment equivalent to residues 159–171 of mouse DNMT1 (mDNMT1) is responsible for interacting with proliferating cell nuclear antigen (PCNA) [71], thereby contributing to the recruitment of DNMT1 to the replication foci during the S phase [71], or the DNA repair sites [72]. The NTD reportedly also interacts with other proteins, including DMAP1 [73], G9a [74], DNMT3A [75], DNMT3B [75], PKC [76] and CDKL [77] to regulate transcription repression, heterochromatin formation or the pathogenic processes of Rett syndrome. In addition, the DNA binding activity of the NTD has been reported [78–80]. However, due to lack of a structural study, the functional implication of most of the NTD-associated interactions remains to be investigated.

2.6. Regulatory Role of DNMT1 (GK)n Repeats

The (GK)n repeat of DNMT1, which is highly conserved throughout evolution, links the regulatory domains to the MTase domain. Current structural studies indicate that this repeat is not involved in the DNA interaction. Rather, it constitutes a binding site for deubiquitinase USP7, an enzyme that plays a regulatory role in DNMT1-mediated maintenance DNA methylation [81–84]. The DNMT1-USP7 interaction is subject to regulation by the acetyltransferase Tip60 and the deacetylase HDAC1: Tip60-mediated acetylation of the (GK)n repeat leads to the disruption of the DNMT1-USP7 interaction, which can be restored by the HDAC1-mediated deacetylation of the same site [82]. On the other hand, a more recent study has suggested that the (GK)n repeat may participate in the DNMT1-mediated de novo methylation of paternal imprinting control regions (ICRs) in mouse ES cells [85]. Due to the lack of molecular details of DNMT1-mediated methylation in cells, the functional implication of the (GK)n repeat remains controversial [86].

3. Structural Basis of DNMT3A-Mediated DNA Methylation

DNMT3A and DNMT3B mediate DNA methylation establishment during gametogenesis and embryogenesis [16,87], and subsequently participate in methylation maintenance [88–90]. The enzymatic activity of DNMT3A/3B in germ cells and embryonic stem cells is further regulated by DNMT3-like (DNMT3L) protein, which lacks DNA methylation activity but functions to stimulate the cofactor binding and enzymatic activity of DNMT3A/3B [7,91–93] and to maintain DNMT3A stability in cells [94]. DNMT3A and DNMT3B are highly related in sequence, both containing a largely disordered NTD, followed by a Pro-Trp-Trp-Pro (PWWP) domain, an Atrx-Dnmt3-Dnmt3l (ADD) domain and a highly homologous MTase domain (Figure 1). DNMT3L contains an N-terminal ADD domain, followed by a MTase-like domain, which is catalytically inactive due to a lack of essential motifs for enzymatic activity (Figure 1) [95,96].

3.1. Enzyme-Substrate Interaction of DNMT3A

The crystal structure of the MTase domain of DNMT3A in complex with the C-terminal domain of DNMT3L (DNMT3L-C) provides the first atomic details of the DNMT3A-DNMT3L complex [97]. The DNMT3A MTase domain forms a tetrameric fold with DNMT3L-C, in the order of 3L-3A-3A-3L, resulting in two DNMT3A-DNMT3L heterodimeric interfaces and one DNMT3A-DNMT3A homodimeric interface. The homodimerization of DNMT3A is mediated by a network of salt bridges and hydrogen bonding interactions, while the heterodimerization of DNMT3A and DNMT3L is mainly driven by hydrophobic stacking interactions between two pairs of phenylalanine residues [97]. Notably, the active sites between the two DNMT3A monomers are separated by ~40 Å, a distance equivalent to one helical turn of DNA. This observation provides the basis for the CpG spacing model, in which the DNMT3A dimer is capable of methylating two CpG sites located across the opposite strands of one DNA duplex, separated by ~10 base-pair (bp) DNA, in one binding event. This model predicts the prevalence of ~10 bp methylation periodicity in cells, which has been supported by a number of biochemical and cellular studies [97,98]. However, the observation that the 10 bp-methylation periodicity also occurs in plants later prompted alternative explanations for the methylation periodicity [99].

Recently, the crystal structure of DNMT3A-DNMT3L in complex with a DNA duplex containing two separate CpG sites (in which the target cytosines are replaced with zebularines [100]) has been determined [101]. The structure reveals a productive state of the DNMT3A-DNA complex, with two CpG/ZpG (Z: zebularine) sites separately targeted by the two DNMT3A monomers of the DNMT3A-DNMT3L tetramer (Figure 6A), therefore confirming the notion of DNMT3A-mediated DNA co-methylation. The structure of the DNA-bound DNMT3A-DNMT3L tetramer resembles that of free DNMT3A-DNMT3L (Figure 6B), with an RMSD of 1.1 Å over 826 aligned Cα atoms. The most notable structural difference arises from a loop from the TRD (TRD loop), which undergoes a disorder-to-order transition upon DNA binding (Figure 6B). The interaction between DNMT3A and DNA is mediated through the catalytic loop, the TRD loop and the DNMT3A-DNMT3A homodimeric interface (Figure 6A), which together create a continuous DNA-binding surface. The zebularines are flipped out of the DNA duplex and insert deep into the catalytic pocket of DNMT3A, where they are covalently anchored by the catalytic cysteine C710 and recognized by several other residues through hydrogen bonding interactions (Figure 6C) [101]. Similar to the productive mDNMT1-DNA complex, the catalytic loop and TRD loop of DNMT3A approach the DNA molecule from the minor groove and the major groove, respectively, with residue V716 from the catalytic loop intercalating into the DNA cavity vacated by base flipping (Figure 6C,D). In the minor groove, the backbone carbonyl of V716 forms a hydrogen bond with the orphan guanine (Figure 6C), while in the major groove, residues R836 and T834 from the TRD loop also interact with the guanine of the target strand through direct and water-mediated hydrogen bonding interactions (Figure 6D). Consistent with these structural observations, the introduction of mutations into these CpG-interacting residues leads to either dramatically decreased activity (for V716G) or altered methylation specificity (for R836A) in vitro and in cells [101]. Mutations of the substrate binding site of DNMT3A, including R882H, have been associated with hematological cancer [102–104]. Both in vitro and in vivo assays indicated that these mutations compromise the enzymatic activity of DNMT3A [101–107], which may contribute to disease progression.

It is worth noting that the structure of the DNMT3A-DNMT3L-DNA complex reveals that the active sites between the two DNMT3A monomers are separated by 14 bp DNA, instead of the 10 bp as previously proposed. Whether this observation arises from the inherent structural property of DNMT3A or its conformational dynamics remains to be investigated.

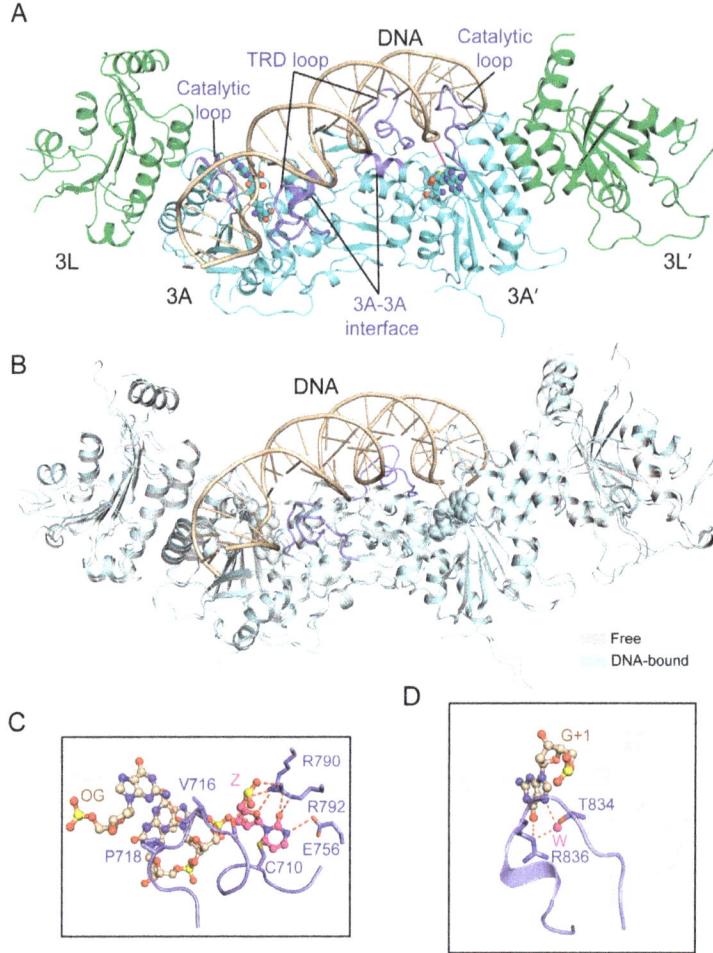

Figure 6. Structural analysis of the productive complex of the DNMT3A-DNMT3L tetramer with CpG DNA. (**A**) Structural overview of the DNMT3A-DNMT3L tetramer covalently bound to a 25-mer DNA duplex containing two CpG/ZpG sites (Z: Zebularine) (PDB 5YX2). The flipped-out zebularines are colored in purple. (**B**) Structural overlap between the DNA-bound and free DNMT3A-DNMT3L tetramer (PDB 2QRV). The TRD loops, which undergo disorder-to-order transition upon DNA binding, are colored in blue. (**C**) The DNA interactions involving the catalytic loop and other catalytic residues. (OG: Orphan guanine). The hydrogen bonding interactions are depicted as dashed lines. (**D**) Residues T834 and R836 from the TRD loop (blue) engage in base-specific recognition of the CpG site. The water molecules are shown as purple spheres.

3.2. ADD Domain-Mediated Autoinhibition of DNMT3A

The ADD domain of DNMT3A is comprised of an N-terminal GATA-like zinc finger, a PHD finger and a C-terminal α-helix [108], together packing into a single globular fold. This domain has been characterized as a reader module that specifically binds to histone H3 unmethylated at lysine 4 (H3K4me0) [108,109]. The association of the DNMT3A ADD domain with H3K4me0 is mediated by antiparallel β-pairing between the two-stranded β-sheet of the ADD domain and residues A1-T6 of H3, with the side chain of H3K4me0 engaging in hydrogen-bonding interactions with D529, D531 and Q534 from the ADD domain [108]. In addition, a downstream loop of the ADD domain undergoes a disorder-to-order transition to close up on the N-terminus of H3, supporting the specific ADD-H3 association [108].

Recent studies have further revealed that the ADD domain regulates the activity of DNMT3A through an H3-dependent, autoinhibitory mechanism [110,111]. The structure of a DNMT3A fragment, spanning the ADD and MTase domains, in complex with DNMT3L-C reveals an intramolecular interaction between the ADD and MTase domains of DNMT3A (Figure 7A). In particular, the linker sequence following the ADD domain initiates a hydrophobic contact with the MTase domain, which then guides the insertion of a loop (residues 526–533) of the ADD domain into the catalytic cleft, where it engages in salt-bridge interactions with DNA binding sites (R790, R792, H789 and R831) (Figure 7B), thereby inhibiting the substrate binding of DNMT3A (Figure 7C) [110]. In contrast, the structure of the DNMT3A-DNMT3L-H3 complex demonstrates that, upon binding to H3 (Figure 7D,E), the DNMT3A ADD domain is repositioned from the catalytic cleft onto a different surface of the MTase domain, engaging a distinct set of hydrogen bonds and hydrophobic interactions (Figure 7D) [110]. The structural comparison of the H3-free and H3-bound DNMT3A complexes therefore provides a dynamic view on how the H3 binding switches the conformation of DNMT3A from an autoinhibitory state to an active state. Note that the residues involved in the autoinhibitory regulation of DNMT3A are highly conserved in DNMT3B, suggesting a conserved allosteric regulation mode of DNMT3 methyltransferases.

The observation that the intramolecular ADD-MTase interaction interplays with the intermolecular ADD-H3 interaction establishes a direct coupling between the enzymatic activity and chromatin targeting of DNMT3A. Similar to the RFTS domain-mediated allosteric regulation of DNMT1, as described above, this regulatory mechanism of the DNMT3A ADD domain ensures the precise spatial regulation of DNMT3A [109–111], which is essential for installing lineage-specific DNA methylation patterns across the genome.

3.3. Functional Regulation of DNMT3A by the N-Terminal Tail and PWWP Domain

The NTD segment defines the most divergent region between DNMT3A and DNMT3B. This region has been shown to regulate the DNA binding and cellular localization of DNMT3A [112–114]. Unlike full-length DNMT3A that is predominantly localized to the heterochromatic region, DNMT3A2, an isoform of DNMT3A lacking residues 1–221 of the NTD, becomes enriched in the euchromatic region, with reduced DNA binding affinity [114]. The precise regulatory role of this domain remains to be investigated.

The PWWP domain, named after a characteristic proline-tryptophan-tryptophan-proline motif, belongs to the Royal super-family of domains that recognize histone tails with various modifications [115,116]. The PWWP domain of DNMT3A and DNMT3B mediates their chromatin association through specific recognition of histone H3 trimethylated at lysine 36 (H3K36me3) [117,118], which is essential for directing the de novo methylation activity of DNMT3A/3B at the pericentric heterochromatin [119]. Structural studies of the DNMT3A/3B PWWP domain revealed a β-barrel followed by a C-terminal helical bundle, similar to other PWWP domains (Figure 8A,B) [120–122]. The β-barrel is comprised of five β-strands, with the signature PWWP motif replaced by a SWWP motif at the beginning of the second β-strand. The structure of the DNMT3B PWWP domain in complex with an H3K36me3 peptide reveals that the histone peptide occupies a surface groove formed by residues from the β1 strand, the β1-β2 loop, and the β4 strand, with the side chain of H3K36me3 inserting into the aromatic cage formed by F236, W239 and W263 through hydrophobic and cation-π interactions (Figure 8B) [121]. The H3K36me3 binding also induces a conformational change of the β1-β2 loop, which moves to close up the aromatic cage, thereby enhancing the specific H3K36me3 recognition. In addition, both the DNMT3A and DNMT3B PWWP domains present a positively charged surface that confers their DNA binding activity (Figure 8C) [120,121,123]. The cooperative engagement of both DNA and H3K36me3 by the DNMT3A/3B PWWP domains provides a mechanism for targeting these two enzymes to heterochromatic regions [118,119] or the actively transcribed gene body in the nucleus [124].

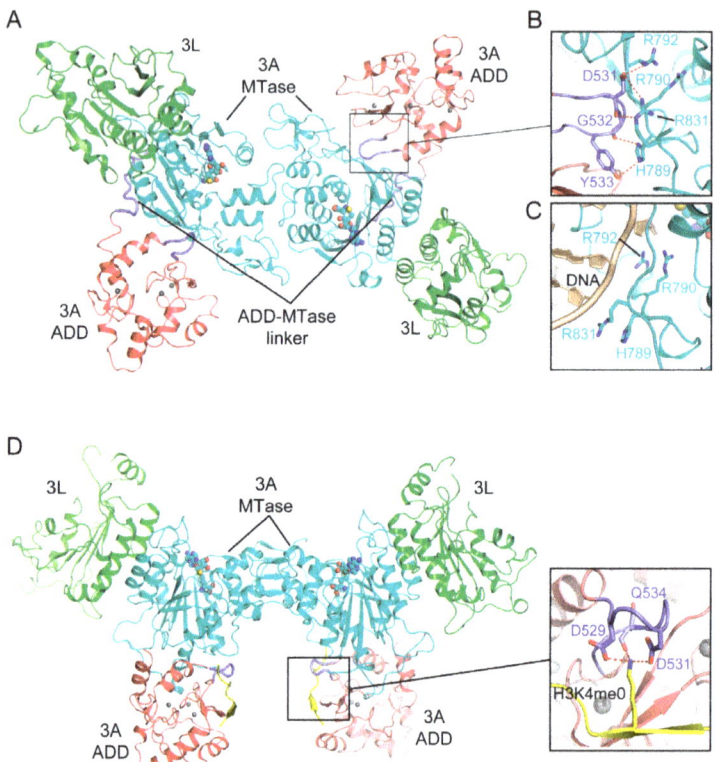

Figure 7. Structural analysis of the Atrx-Dnmt3-Dnmt3l (ADD) domain-mediated DNMT3A autoinhibition. (**A**) Structural overview of the DNMT3A-DNMT3L tetramer, with the DNMT3A fragment comprised of both the ADD and MTase domains (PDB 4U7P). (**B**) Intramolecular interactions between the ADD loop (blue) and the MTase domain (aquamarine) of DNMT3A. The hydrogen bonding interactions are depicted as dashed lines. (**C**) The ADD-binding site of the DNMT3A MTase overlaps with its DNA binding site. (**D**) Structure of the DNMT3A-DNMT3L tetramer bound to the histone H3K4me0 peptide (PDB 4U7T), with the interaction between H3K4me0 and the ADD domain shown in an expanded view (PDB 3A1B).

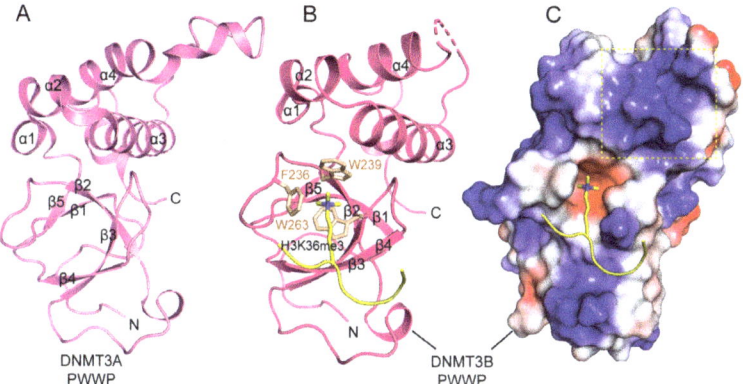

Figure 8. Structures of DNMT3A/3B Pro-Trp-Trp-Pro (PWWP) domains. (**A**) Crystal structure of the DNMT3A PWWP domain (PDB 3LLR). (**B**) Crystal structure of DNMT3B PWWP bound to the histone H3K36me3 peptide (PDB 5CIU), with the side chains of H3K36me3 and the cage residues of the PWWP domain shown in stick representation. (**C**) Electrostatic surface view of the DNMT3B PWWP domain bound to the histone H3K36me3 peptide. The putative DNA binding surface is boxed with dotted lines.

4. Structural Comparison of the DNMT1-DNA and DNMT3A-DNA Complexes

The structural comparison of the DNMT3A-DNMT3L-DNA complex and the mDNMT1-hemimethylated DNA complex provides insights into the distinct molecular basis between DNMT3A-mediated de novo DNA methylation and DNMT1-mediated maintenance DNA methylation. Despite the conformational similarity in their catalytic loop for accessing the DNA minor groove, mDNMT1 and DNMT3A enter the DNA major groove differently for CpG recognition (Figure 9A–D). Firstly, mDNMT1 interacts with the DNA major groove through two of its TRD loops, with one (TRD loop 1) engaging the CpG dinucleotide through hydrogen bonding interactions and the other (TRD loop 2) forming a hydrophobic concave harboring the methyl group of 5mC along the template strand (Figure 9A,B). In contrast, while DNMT3A interacts with the DNA major groove through a loop similar to TRD loop 1 in DNMT1, it lacks the DNMT1 TRD loop 2-equivalent segment for 5mC recognition (Figure 9C,D). These observations explain why DNMT1, but not DNMT3A, shows an enzymatic preference for hemimethylated substrates over unmethylated substrates. Additionally, the DNA molecules bound to mDNMT1 and DNMT3A also exhibit different conformational adjustments. In mDNMT1-bound DNA, the base flipping leads to one-base translocation of the orphan guanine and a large distortion of the CpG site, with the DNA cavity filled by two bulky protein residues (M1235 and K1537) (Figure 9A). In contrast, in DNMT3A-bound DNA, the orphan guanine remains in space, resulting in a smaller DNA cavity occupied by one small residue of DNMT3A (V716) (Figure 9C). In addition, the large TRD of DNMT1 permits an extensive protein-DNA interaction, resulting in a buried surface area of ~2100 $Å^2$, whereas the DNA binding of DNMT3A, with a much smaller TRD, only leads to buried surface area of ~1300 $Å^2$ for each DNMT3A monomer. This limited DNA binding of each DNMT3A monomer is nevertheless overcome by the presence of two DNMT3A monomers in the DNMT3A-DNMT3L tetramer, which provides an enlarged protein-DNA contact surface to ensure the efficiency of DNA methylation. Together, these observations highlight the molecular basis underlying the difference between DNMT3A-mediated de novo methylation and DNMT1-mediated maintenance methylation.

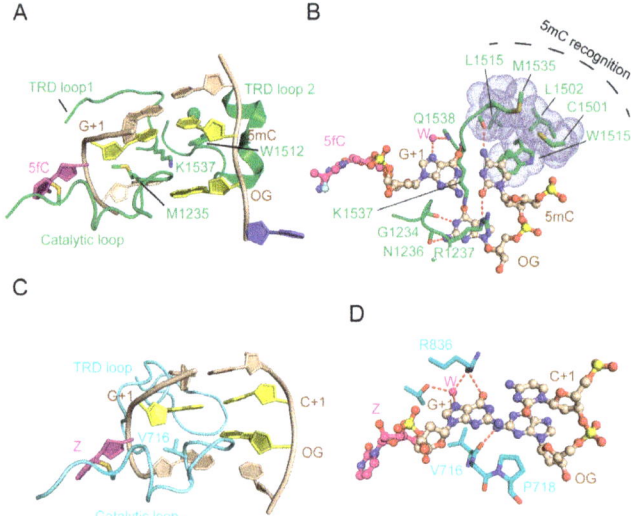

Figure 9. Structural comparison of mDNMT1-DNA and DNMT3A-DNA interactions. (**A**) Recognition of hemimethylated CpG DNA by mDNMT1 (green) (PDB 4DA4). The hemimethylated CpG site, containing a 5-methyl group (green sphere), is colored in yellow or purple (5-fluorocytosine, 5fC′). The flipped-out cytosine on the template strand is colored in blue and the rest of the DNA is colored in wheat. (**B**) Detailed interactions between mDNMT1 and the hemimethylated CpG site. (**C**) Recognition of unmodified CpG DNA by DNMT3A (light blue). (**D**) Detailed interactions between DNMT3A and the unmodified CpG site. The hydrogen bonds are depicted as dashed lines.

5. Summary

Recent structural and biochemical studies have greatly advanced our understanding of DNMT1-mediated maintenance DNA methylation and DNMT3A/3B-mediated de novo DNA methylation. Structural elucidations of DNMT1 and DNMT3A in complex with their respective DNA substrates or histone peptides provide mechanistic details for the functional regulation and substrate specificity of these enzymes. However, a number of outstanding questions remain to be addressed, for example, how are the N-terminal domains of DNMT1 or DNMT3A coordinated in regulating the enzymatic activity and genome targeting? How are the DNMTs regulated in the chromatin environment? Future investigations of the structure and dynamics of DNMT1 and DNMT3A/3B in their cellular environment will help provide a systematic view on the mechanistic basis of mammalian DNA methylation.

Funding: This research was funded by NIH (1R35GM119721 to J.S.).

Conflicts of Interest: The authors declare no conflict of interest.

References

1. Razin, A.; Riggs, A.D. DNA methylation and gene function. *Science* **1980**, *210*, 604–610. [CrossRef] [PubMed]
2. Law, J.A.; Jacobsen, S.E. Establishing, maintaining and modifying DNA methylation patterns in plants and animals. *Nat. Rev. Genet.* **2010**, *11*, 204–220. [CrossRef]
3. Razin, A.; Cedar, H. DNA methylation and gene expression. *Microbiol. Rev.* **1991**, *55*, 451–458.
4. Siegfried, Z.; Cedar, H. DNA methylation: A molecular lock. *Curr. Biol.* **1997**, *7*, R305–R307. [CrossRef]
5. Bird, A.P.; Wolffe, A.P. Methylation-induced repression–belts, braces, and chromatin. *Cell* **1999**, *99*, 451–454. [CrossRef]

6. Smith, Z.D.; Meissner, A. DNA methylation: Roles in mammalian development. *Nat. Rev. Genet.* **2013**, *14*, 204–220. [CrossRef]
7. Bourc'his, D.; Bestor, T.H. Meiotic catastrophe and retrotransposon reactivation in male germ cells lacking Dnmt3L. *Nature* **2004**, *431*, 96–99. [CrossRef]
8. Holliday, R.; Pugh, J.E. DNA modification mechanisms and gene activity during development. *Science* **1975**, *187*, 226–232. [CrossRef]
9. Walsh, C.P.; Chaillet, J.R.; Bestor, T.H. Transcription of IAP endogenous retroviruses is constrained by cytosine methylation. *Nat. Genet.* **1998**, *20*, 116–117. [CrossRef]
10. Li, E.; Beard, C.; Jaenisch, R. Role for DNA methylation in genomic imprinting. *Nature* **1993**, *366*, 362–365. [CrossRef]
11. Stoger, R.; Kubicka, P.; Liu, C.G.; Kafri, T.; Razin, A.; Cedar, H.; Barlow, D.P. Maternal-specific methylation of the imprinted mouse Igf2r locus identifies the expressed locus as carrying the imprinting signal. *Cell* **1993**, *73*, 61–71. [CrossRef]
12. Panning, B.; Jaenisch, R. RNA and the epigenetic regulation of X chromosome inactivation. *Cell* **1998**, *93*, 305–308. [CrossRef]
13. Riggs, A.D. X inactivation, differentiation, and DNA methylation. *Cytogenet. Cell Genet.* **1975**, *14*, 9–25. [CrossRef]
14. Ehrlich, M.; Gama-Sosa, M.A.; Huang, L.H.; Midgett, R.M.; Kuo, K.C.; McCune, R.A.; Gehrke, C. Amount and distribution of 5-methylcytosine in human DNA from different types of tissues of cells. *Nucleic Acids Res.* **1982**, *10*, 2709–2721. [CrossRef] [PubMed]
15. Kato, Y.; Kaneda, M.; Hata, K.; Kumaki, K.; Hisano, M.; Kohara, Y.; Okano, M.; Li, E.; Nozaki, M.; Sasaki, H. Role of the Dnmt3 family in de novo methylation of imprinted and repetitive sequences during male germ cell development in the mouse. *Hum. Mol. Genet.* **2007**, *16*, 2272–2280. [CrossRef] [PubMed]
16. Okano, M.; Bell, D.W.; Haber, D.A.; Li, E. DNA methyltransferases Dnmt3a and Dnmt3b are essential for de novo methylation and mammalian development. *Cell* **1999**, *99*, 247–257. [CrossRef]
17. Chen, T.; Li, E. Structure and function of eukaryotic DNA methyltransferases. *Curr. Top. Dev. Biol.* **2004**, *60*, 55–89. [PubMed]
18. Goll, M.G.; Bestor, T.H. Eukaryotic cytosine methyltransferases. *Annu Rev. Biochem.* **2005**, *74*, 481–514. [CrossRef]
19. Dodge, J.E.; Okano, M.; Dick, F.; Tsujimoto, N.; Chen, T.; Wang, S.; Ueda, Y.; Dyson, N.; Li, E. Inactivation of Dnmt3b in mouse embryonic fibroblasts results in DNA hypomethylation, chromosomal instability, and spontaneous immortalization. *J. Biol. Chem.* **2005**, *280*, 17986–17991. [CrossRef]
20. Feng, J.; Zhou, Y.; Campbell, S.L.; Le, T.; Li, E.; Sweatt, J.D.; Silva, A.J.; Fan, G. Dnmt1 and Dnmt3a maintain DNA methylation and regulate synaptic function in adult forebrain neurons. *Nat. Neurosci.* **2010**, *13*, 423–430. [CrossRef]
21. Feltus, F.A.; Lee, E.K.; Costello, J.F.; Plass, C.; Vertino, P.M. Predicting aberrant CpG island methylation. *Proc. Natl. Acad. Sci. USA* **2003**, *100*, 12253–12258. [CrossRef] [PubMed]
22. Jair, K.W.; Bachman, K.E.; Suzuki, H.; Ting, A.H.; Rhee, I.; Yen, R.W.; Baylin, S.B.; Schuebel, K.E. De novo CpG island methylation in human cancer cells. *Cancer Res.* **2006**, *66*, 682–692. [CrossRef] [PubMed]
23. Schubert, H.L.; Blumenthal, R.M.; Cheng, X. Many paths to methyltransfer: A chronicle of convergence. *Trends Biochem. Sci.* **2003**, *28*, 329–335. [CrossRef]
24. Martin, J.L.; McMillan, F.M. SAM (dependent) I AM: The S-adenosylmethionine-dependent methyltransferase fold. *Curr. Opin. Struct. Biol.* **2002**, *12*, 783–793. [CrossRef]
25. Denis, H.; Ndlovu, M.N.; Fuks, F. Regulation of mammalian DNA methyltransferases: A route to new mechanisms. *EMBO Rep.* **2011**, *12*, 647–656. [CrossRef] [PubMed]
26. Jeltsch, A.; Jurkowska, R.Z. Allosteric control of mammalian DNA methyltransferases—A new regulatory paradigm. *Nucleic Acids Res.* **2016**, *44*, 8556–8575. [CrossRef] [PubMed]
27. Cheng, X.; Blumenthal, R.M. Mammalian DNA methyltransferases: A structural perspective. *Structure* **2008**, *16*, 341–350. [CrossRef] [PubMed]
28. Jeltsch, A. On the enzymatic properties of Dnmt1: Specificity, processivity, mechanism of linear diffusion and allosteric regulation of the enzyme. *Epigenetics* **2006**, *1*, 63–66. [CrossRef]
29. Bird, A.P. Use of restriction enzymes to study eukaryotic DNA methylation: II. The symmetry of methylated sites supports semi-conservative copying of the methylation pattern. *J. Mol. Biol.* **1978**, *118*, 49–60. [CrossRef]

30. Song, J.; Rechkoblit, O.; Bestor, T.H.; Patel, D.J. Structure of DNMT1-DNA complex reveals a role for autoinhibition in maintenance DNA methylation. *Science* **2011**, *331*, 1036–1040. [CrossRef]
31. Song, J.; Teplova, M.; Ishibe-Murakami, S.; Patel, D.J. Structure-based mechanistic insights into DNMT1-mediated maintenance DNA methylation. *Science* **2012**, *335*, 709–712. [CrossRef] [PubMed]
32. Takeshita, K.; Suetake, I.; Yamashita, E.; Suga, M.; Narita, H.; Nakagawa, A.; Tajima, S. Structural insight into maintenance methylation by mouse DNA methyltransferase 1 (Dnmt1). *Proc. Natl. Acad. Sci. USA* **2011**, *108*, 9055–9059. [CrossRef] [PubMed]
33. Zhang, Z.M.; Liu, S.; Lin, K.; Luo, Y.; Perry, J.J.; Wang, Y.; Song, J. Crystal Structure of Human DNA Methyltransferase 1. *J. Mol. Biol.* **2015**, *427*, 2520–2531. [CrossRef] [PubMed]
34. Osterman, D.G.; DePillis, G.D.; Wu, J.C.; Matsuda, A.; Santi, D.V. 5-Fluorocytosine in DNA is a mechanism-based inhibitor of HhaI methylase. *Biochemistry* **1988**, *27*, 5204–5210. [CrossRef] [PubMed]
35. Yang, N.; Xu, R.M. Structure and function of the BAH domain in chromatin biology. *Crit. Rev. Biochem. Mol. Biol.* **2013**, *48*, 211–221. [CrossRef] [PubMed]
36. Hermann, A.; Goyal, R.; Jeltsch, A. The Dnmt1 DNA-(cytosine-C5)-methyltransferase methylates DNA processively with high preference for hemimethylated target sites. *J. Biol. Chem.* **2004**, *279*, 48350–48359. [CrossRef]
37. Vilkaitis, G.; Suetake, I.; Klimasauskas, S.; Tajima, S. Processive methylation of hemimethylated CpG sites by mouse Dnmt1 DNA methyltransferase. *J. Biol. Chem.* **2005**, *280*, 64–72. [CrossRef]
38. Pradhan, M.; Esteve, P.O.; Chin, H.G.; Samaranayke, M.; Kim, G.D.; Pradhan, S. CXXC domain of human DNMT1 is essential for enzymatic activity. *Biochemistry* **2008**, *47*, 10000–10009. [CrossRef]
39. Ye, F.; Kong, X.; Zhang, H.; Liu, Y.; Shao, Z.; Jin, J.; Cai, Y.; Zhang, R.; Li, L.; Zhang, Y.W.; et al. Biochemical Studies and molecular dynamic simulations reveal the molecular basis of conformational changes in DNA methyltransferase-1. *ACS Chem. Biol.* **2018**, *13*, 772–781. [CrossRef]
40. Bashtrykov, P.; Jankevicius, G.; Smarandache, A.; Jurkowska, R.Z.; Ragozin, S.; Jeltsch, A. Specificity of Dnmt1 for methylation of hemimethylated CpG sites resides in its catalytic domain. *Chem. Biol.* **2012**, *19*, 572–578. [CrossRef]
41. Jurkowska, R.Z.; Jurkowski, T.P.; Jeltsch, A. Structure and function of mammalian DNA methyltransferases. *Chembiochem* **2011**, *12*, 206–222. [CrossRef] [PubMed]
42. Easwaran, H.P.; Schermelleh, L.; Leonhardt, H.; Cardoso, M.C. Replication-independent chromatin loading of Dnmt1 during G2 and M phases. *EMBO Rep.* **2004**, *5*, 1181–1186. [CrossRef] [PubMed]
43. Leonhardt, H.; Page, A.W.; Weier, H.U.; Bestor, T.H. A targeting sequence directs DNA methyltransferase to sites of DNA replication in mammalian nuclei. *Cell* **1992**, *71*, 865–873. [CrossRef]
44. Klein, C.J.; Bird, T.; Ertekin-Taner, N.; Lincoln, S.; Hjorth, R.; Wu, Y.; Kwok, J.; Mer, G.; Dyck, P.J.; Nicholson, G.A. DNMT1 mutation hot spot causes varied phenotypes of HSAN1 with dementia and hearing loss. *Neurology* **2013**, *80*, 824–828. [CrossRef] [PubMed]
45. Klein, C.J.; Botuyan, M.V.; Wu, Y.; Ward, C.J.; Nicholson, G.A.; Hammans, S.; Hojo, K.; Yamanishi, H.; Karpf, A.R.; Wallace, D.C.; et al. Mutations in DNMT1 cause hereditary sensory neuropathy with dementia and hearing loss. *Nat. Genet.* **2011**, *43*, 595–600. [CrossRef] [PubMed]
46. Pedroso, J.L.; Povoas Barsottini, O.G.; Lin, L.; Melberg, A.; Oliveira, A.S.; Mignot, E. A novel de novo exon 21 DNMT1 mutation causes cerebellar ataxia, deafness, and narcolepsy in a Brazilian patient. *Sleep* **2013**, *36*, 1257–1259, 1259A. [CrossRef] [PubMed]
47. Winkelmann, J.; Lin, L.; Schormair, B.; Kornum, B.R.; Faraco, J.; Plazzi, G.; Melberg, A.; Cornelio, F.; Urban, A.E.; Pizza, F.; et al. Mutations in DNMT1 cause autosomal dominant cerebellar ataxia, deafness and narcolepsy. *Hum. Mol. Genet.* **2012**, *21*, 2205–2210. [CrossRef] [PubMed]
48. Ishiyama, S.; Nishiyama, A.; Saeki, Y.; Moritsugu, K.; Morimoto, D.; Yamaguchi, L.; Arai, N.; Matsumura, R.; Kawakami, T.; Mishima, Y.; et al. Structure of the Dnmt1 reader module complexed with a unique two-mono-ubiquitin mark on histone H3 reveals the basis for DNA methylation maintenance. *Mol. Cell* **2017**, *68*, 350–360 e357. [CrossRef]
49. Nishiyama, A.; Yamaguchi, L.; Sharif, J.; Johmura, Y.; Kawamura, T.; Nakanishi, K.; Shimamura, S.; Arita, K.; Kodama, T.; Ishikawa, F.; et al. Uhrf1-dependent H3K23 ubiquitylation couples maintenance DNA methylation and replication. *Nature* **2013**, *502*, 249–253. [CrossRef]

50. Qin, W.; Wolf, P.; Liu, N.; Link, S.; Smets, M.; La Mastra, F.; Forne, I.; Pichler, G.; Horl, D.; Fellinger, K.; et al. DNA methylation requires a DNMT1 ubiquitin interacting motif (UIM) and histone ubiquitination. *Cell Res.* **2015**, *25*, 911–929. [CrossRef]
51. Bostick, M.; Kim, J.K.; Esteve, P.O.; Clark, A.; Pradhan, S.; Jacobsen, S.E. UHRF1 plays a role in maintaining DNA methylation in mammalian cells. *Science* **2007**, *317*, 1760–1764. [CrossRef] [PubMed]
52. Sharif, J.; Muto, M.; Takebayashi, S.; Suetake, I.; Iwamatsu, A.; Endo, T.A.; Shinga, J.; Mizutani-Koseki, Y.; Toyoda, T.; Okamura, K.; et al. The SRA protein Np95 mediates epigenetic inheritance by recruiting Dnmt1 to methylated DNA. *Nature* **2007**, *450*, 908–912. [CrossRef] [PubMed]
53. Hashimoto, H.; Horton, J.R.; Zhang, X.; Cheng, X. UHRF1, a modular multi-domain protein, regulates replication-coupled crosstalk between DNA methylation and histone modifications. *Epigenetics* **2009**, *4*, 8–14. [CrossRef] [PubMed]
54. Fang, J.; Cheng, J.; Wang, J.; Zhang, Q.; Liu, M.; Gong, R.; Wang, P.; Zhang, X.; Feng, Y.; Lan, W.; et al. Hemi-methylated DNA opens a closed conformation of UHRF1 to facilitate its histone recognition. *Nat. Commun.* **2016**, *7*, 11197. [CrossRef] [PubMed]
55. Gao, L.; Tan, X.F.; Zhang, S.; Wu, T.; Zhang, Z.M.; Ai, H.W.; Song, J. An intramolecular interaction of UHRF1 reveals dual control for its histone association. *Structure* **2018**, *26*, 304–311. [CrossRef] [PubMed]
56. Gelato, K.A.; Tauber, M.; Ong, M.S.; Winter, S.; Hiragami-Hamada, K.; Sindlinger, J.; Lemak, A.; Bultsma, Y.; Houliston, S.; Schwarzer, D.; et al. Accessibility of different histone H3-binding domains of UHRF1 is allosterically regulated by phosphatidylinositol 5-phosphate. *Mol. Cell* **2014**, *54*, 905–919. [CrossRef] [PubMed]
57. Zhang, Z.M.; Rothbart, S.B.; Allison, D.F.; Cai, Q.; Harrison, J.S.; Li, L.; Wang, Y.; Strahl, B.D.; Wang, G.G.; Song, J. An allosteric interaction links USP7 to Deubiquitination and chromatin targeting of UHRF1. *Cell Rep.* **2015**, *12*, 1400–1406. [CrossRef]
58. Arita, K.; Isogai, S.; Oda, T.; Unoki, M.; Sugita, K.; Sekiyama, N.; Kuwata, K.; Hamamoto, R.; Tochio, H.; Sato, M.; et al. Recognition of modification status on a histone H3 tail by linked histone reader modules of the epigenetic regulator UHRF1. *Proc. Natl. Acad. Sci. USA* **2012**, *109*, 12950–12955. [CrossRef]
59. Cheng, J.; Yang, Y.; Fang, J.; Xiao, J.; Zhu, T.; Chen, F.; Wang, P.; Li, Z.; Yang, H.; Xu, Y. Structural insight into coordinated recognition of trimethylated histone H3 lysine 9 (H3K9me3) by the plant homeodomain (PHD) and tandem tudor domain (TTD) of UHRF1 (ubiquitin-like, containing PHD and RING finger domains, 1) protein. *J. Biol. Chem.* **2013**, *288*, 1329–1339. [CrossRef]
60. Karagianni, P.; Amazit, L.; Qin, J.; Wong, J. ICBP90, a novel methyl K9 H3 binding protein linking protein ubiquitination with heterochromatin formation. *Mol. Cell. Biol.* **2008**, *28*, 705–717. [CrossRef]
61. Liu, X.; Gao, Q.; Li, P.; Zhao, Q.; Zhang, J.; Li, J.; Koseki, H.; Wong, J. UHRF1 targets DNMT1 for DNA methylation through cooperative binding of hemi-methylated DNA and methylated H3K9. *Nat. Commun.* **2013**, *4*, 1563. [CrossRef]
62. Rothbart, S.B.; Dickson, B.M.; Ong, M.S.; Krajewski, K.; Houliston, S.; Kireev, D.B.; Arrowsmith, C.H.; Strahl, B.D. Multivalent histone engagement by the linked tandem Tudor and PHD domains of UHRF1 is required for the epigenetic inheritance of DNA methylation. *Genes Dev.* **2013**, *27*, 1288–1298. [CrossRef]
63. Rothbart, S.B.; Krajewski, K.; Nady, N.; Tempel, W.; Xue, S.; Badeaux, A.I.; Barsyte-Lovejoy, D.; Martinez, J.Y.; Bedford, M.T.; Fuchs, S.M.; et al. Association of UHRF1 with methylated H3K9 directs the maintenance of DNA methylation. *Nat. Struct. Mol. Biol.* **2012**, *19*, 1155–1160. [CrossRef] [PubMed]
64. Xie, S.; Jakoncic, J.; Qian, C. UHRF1 double tudor domain and the adjacent PHD finger act together to recognize K9me3-containing histone H3 tail. *J. Mol. Biol.* **2012**, *415*, 318–328. [CrossRef]
65. Kouzarides, T. Chromatin modifications and their function. *Cell* **2007**, *128*, 693–705. [CrossRef] [PubMed]
66. Arita, K.; Ariyoshi, M.; Tochio, H.; Nakamura, Y.; Shirakawa, M. Recognition of hemi-methylated DNA by the SRA protein UHRF1 by a base-flipping mechanism. *Nature* **2008**, *455*, 818–821. [CrossRef]
67. Avvakumov, G.V.; Walker, J.R.; Xue, S.; Li, Y.; Duan, S.; Bronner, C.; Arrowsmith, C.H.; Dhe-Paganon, S. Structural basis for recognition of hemi-methylated DNA by the SRA domain of human UHRF1. *Nature* **2008**, *455*, 822–825. [CrossRef]
68. Hashimoto, H.; Horton, J.R.; Zhang, X.; Bostick, M.; Jacobsen, S.E.; Cheng, X. The SRA domain of UHRF1 flips 5-methylcytosine out of the DNA helix. *Nature* **2008**, *455*, 826–829. [CrossRef]
69. Sharif, J.; Koseki, H. Recruitment of Dnmt1 roles of the SRA protein Np95 (Uhrf1) and other factors. *Prog. Mol. Biol. Transl. Sci.* **2011**, *101*, 289–310.

70. Harrison, J.S.; Cornett, E.M.; Goldfarb, D.; DaRosa, P.A.; Li, Z.M.; Yan, F.; Dickson, B.M.; Guo, A.H.; Cantu, D.V.; Kaustov, L.; et al. Hemi-methylated DNA regulates DNA methylation inheritance through allosteric activation of H3 ubiquitylation by UHRF1. *eLife* **2016**, *5*, e17101. [CrossRef] [PubMed]
71. Chuang, L.S.; Ian, H.I.; Koh, T.W.; Ng, H.H.; Xu, G.; Li, B.F. Human DNA-(cytosine-5) methyltransferase-PCNA complex as a target for p21WAF1. *Science* **1997**, *277*, 1996–2000. [CrossRef] [PubMed]
72. Mortusewicz, O.; Schermelleh, L.; Walter, J.; Cardoso, M.C.; Leonhardt, H. Recruitment of DNA methyltransferase I to DNA repair sites. *Proc. Natl. Acad. Sci. USA* **2005**, *102*, 8905–8909. [CrossRef]
73. Rountree, M.R.; Bachman, K.E.; Baylin, S.B. DNMT1 binds HDAC2 and a new co-repressor, DMAP1, to form a complex at replication foci. *Nat. Genet.* **2000**, *25*, 269–277. [CrossRef] [PubMed]
74. Esteve, P.O.; Chin, H.G.; Smallwood, A.; Feehery, G.R.; Gangisetty, O.; Karpf, A.R.; Carey, M.F.; Pradhan, S. Direct interaction between DNMT1 and G9a coordinates DNA and histone methylation during replication. *Genes Dev.* **2006**, *20*, 3089–3103. [CrossRef]
75. Kim, G.D.; Ni, J.; Kelesoglu, N.; Roberts, R.J.; Pradhan, S. Co-operation and communication between the human maintenance and de novo DNA (cytosine-5) methyltransferases. *EMBO J.* **2002**, *21*, 4183–4195. [CrossRef] [PubMed]
76. Lavoie, G.; Esteve, P.O.; Laulan, N.B.; Pradhan, S.; St-Pierre, Y. PKC isoforms interact with and phosphorylate DNMT1. *BMC Biol.* **2011**, *9*, 31. [CrossRef] [PubMed]
77. Kameshita, I.; Sekiguchi, M.; Hamasaki, D.; Sugiyama, Y.; Hatano, N.; Suetake, I.; Tajima, S.; Sueyoshi, N. Cyclin-dependent kinase-like 5 binds and phosphorylates DNA methyltransferase 1. *Biochem. Biophys. Res. Commun.* **2008**, *377*, 1162–1167. [CrossRef] [PubMed]
78. Araujo, F.D.; Croteau, S.; Slack, A.D.; Milutinovic, S.; Bigey, P.; Price, G.B.; Zannis-Hadjopoulos, M.; Szyf, M. The DNMT1 target recognition domain resides in the N terminus. *J. Biol. Chem.* **2001**, *276*, 6930–6936. [CrossRef] [PubMed]
79. Fatemi, M.; Hermann, A.; Pradhan, S.; Jeltsch, A. The activity of the murine DNA methyltransferase Dnmt1 is controlled by interaction of the catalytic domain with the N-terminal part of the enzyme leading to an allosteric activation of the enzyme after binding to methylated DNA. *J. Mol. Biol.* **2001**, *309*, 1189–1199. [CrossRef] [PubMed]
80. Suetake, I.; Hayata, D.; Tajima, S. The amino-terminus of mouse DNA methyltransferase 1 forms an independent domain and binds to DNA with the sequence involving PCNA binding motif. *J. Biochem.* **2006**, *140*, 763–776. [CrossRef]
81. Cheng, J.; Yang, H.; Fang, J.; Ma, L.; Gong, R.; Wang, P.; Li, Z.; Xu, Y. Molecular mechanism for USP7-mediated DNMT1 stabilization by acetylation. *Nat. Commun.* **2015**, *6*, 7023. [CrossRef] [PubMed]
82. Du, Z.; Song, J.; Wang, Y.; Zhao, Y.; Guda, K.; Yang, S.; Kao, H.Y.; Xu, Y.; Willis, J.; Markowitz, S.D.; et al. DNMT1 stability is regulated by proteins coordinating deubiquitination and acetylation-driven ubiquitination. *Sci. Signal.* **2010**, *3*, ra80. [CrossRef] [PubMed]
83. Felle, M.; Joppien, S.; Nemeth, A.; Diermeier, S.; Thalhammer, V.; Dobner, T.; Kremmer, E.; Kappler, R.; Langst, G. The USP7/Dnmt1 complex stimulates the DNA methylation activity of Dnmt1 and regulates the stability of UHRF1. *Nucleic Acids Res.* **2011**, *39*, 8355–8365. [CrossRef] [PubMed]
84. Qin, W.; Leonhardt, H.; Spada, F. Usp7 and Uhrf1 control ubiquitination and stability of the maintenance DNA methyltransferase Dnmt1. *J. Cell. Biochem.* **2011**, *112*, 439–444. [CrossRef] [PubMed]
85. Yarychkivska, O.; Shahabuddin, Z.; Comfort, N.; Boulard, M.; Bestor, T.H. BAH domains and a histone-like motif in DNA methyltransferase 1 (DNMT1) regulate de novo and maintenance methylation in vivo. *J. Biol. Chem.* **2018**. [CrossRef] [PubMed]
86. Yarychkivska, O.; Tavana, O.; Gu, W.; Bestor, T.H. Independent functions of DNMT1 and USP7 at replication foci. *Epigenet. Chromatin* **2018**, *11*, 9. [CrossRef] [PubMed]
87. Okano, M.; Xie, S.; Li, E. Cloning and characterization of a family of novel mammalian DNA (cytosine-5) methyltransferases. *Nat. Genet.* **1998**, *19*, 219–220. [CrossRef]
88. Jeltsch, A.; Jurkowska, R.Z. New concepts in DNA methylation. *Trends Biochem. Sci.* **2014**, *39*, 310–318. [CrossRef]
89. Jones, P.A.; Liang, G. Rethinking how DNA methylation patterns are maintained. *Nat. Rev. Genet.* **2009**, *10*, 805–811. [CrossRef]

90. Riggs, A.D.; Xiong, Z. Methylation and epigenetic fidelity. *Proc. Natl. Acad. Sci. USA* **2004**, *101*, 4–5. [CrossRef]
91. Bourc'his, D.; Xu, G.L.; Lin, C.S.; Bollman, B.; Bestor, T.H. Dnmt3L and the establishment of maternal genomic imprints. *Science* **2001**, *294*, 2536–2539. [CrossRef]
92. Chedin, F.; Lieber, M.R.; Hsieh, C.L. The DNA methyltransferase-like protein DNMT3L stimulates de novo methylation by Dnmt3a. *Proc. Natl. Acad. Sci. USA* **2002**, *99*, 16916–16921. [CrossRef] [PubMed]
93. Hata, K.; Okano, M.; Lei, H.; Li, E. Dnmt3L cooperates with the Dnmt3 family of de novo DNA methyltransferases to establish maternal imprints in mice. *Development* **2002**, *129*, 1983–1993. [PubMed]
94. Veland, N.; Lu, Y.; Hardikar, S.; Gaddis, S.; Zeng, Y.; Liu, B.; Estecio, M.R.; Takata, Y.; Lin, K.; Tomida, M.W.; et al. DNMT3L facilitates DNA methylation partly by maintaining DNMT3A stability in mouse embryonic stem cells. *Nucleic Acids Res.* **2018**. [CrossRef] [PubMed]
95. Aapola, U.; Kawasaki, K.; Scott, H.S.; Ollila, J.; Vihinen, M.; Heino, M.; Shintani, A.; Minoshima, S.; Krohn, K.; Antonarakis, S.E.; et al. Isolation and initial characterization of a novel zinc finger gene, *DNMT3L*, on 21q22.3, related to the cytosine-5-methyltransferase 3 gene family. *Genomics* **2000**, *65*, 293–298. [CrossRef] [PubMed]
96. Ooi, S.K.; Qiu, C.; Bernstein, E.; Li, K.; Jia, D.; Yang, Z.; Erdjument-Bromage, H.; Tempst, P.; Lin, S.P.; Allis, C.D.; et al. DNMT3L connects unmethylated lysine 4 of histone H3 to de novo methylation of DNA. *Nature* **2007**, *448*, 714–717. [CrossRef] [PubMed]
97. Jia, D.; Jurkowska, R.Z.; Zhang, X.; Jeltsch, A.; Cheng, X. Structure of Dnmt3a bound to Dnmt3L suggests a model for de novo DNA methylation. *Nature* **2007**, *449*, 248–251. [CrossRef] [PubMed]
98. Jurkowska, R.Z.; Anspach, N.; Urbanke, C.; Jia, D.; Reinhardt, R.; Nellen, W.; Cheng, X.; Jeltsch, A. Formation of nucleoprotein filaments by mammalian DNA methyltransferase Dnmt3a in complex with regulator Dnmt3L. *Nucleic Acids Res.* **2008**, *36*, 6656–6663. [CrossRef] [PubMed]
99. Chodavarapu, R.K.; Feng, S.; Bernatavichute, Y.V.; Chen, P.Y.; Stroud, H.; Yu, Y.; Hetzel, J.A.; Kuo, F.; Kim, J.; Cokus, S.J.; et al. Relationship between nucleosome positioning and DNA methylation. *Nature* **2010**, *466*, 388–392. [CrossRef]
100. Zhou, L.; Cheng, X.; Connolly, B.A.; Dickman, M.J.; Hurd, P.J.; Hornby, D.P. Zebularine: A novel DNA methylation inhibitor that forms a covalent complex with DNA methyltransferases. *J. Mol. Biol.* **2002**, *321*, 591–599. [CrossRef]
101. Zhang, Z.M.; Lu, R.; Wang, P.; Yu, Y.; Chen, D.; Gao, L.; Liu, S.; Ji, D.; Rothbart, S.B.; Wang, Y.; et al. Structural basis for DNMT3A-mediated de novo DNA methylation. *Nature* **2018**, *554*, 387–391. [CrossRef] [PubMed]
102. Robertson, K.D. DNA methylation and human disease. *Nat. Rev. Genet.* **2005**, *6*, 597–610. [CrossRef]
103. Yang, L.; Rau, R.; Goodell, M.A. DNMT3A in haematological malignancies. *Nat. Rev. Cancer* **2015**, *15*, 152–165. [CrossRef]
104. Ley, T.J.; Ding, L.; Walter, M.J.; McLellan, M.D.; Lamprecht, T.; Larson, D.E.; Kandoth, C.; Payton, J.E.; Baty, J.; Welch, J.; et al. DNMT3A mutations in acute myeloid leukemia. *N. Engl. J. Med.* **2010**, *363*, 2424–2433. [CrossRef] [PubMed]
105. Emperle, M.; Dukatz, M.; Kunert, S.; Holzer, K.; Rajavelu, A.; Jurkowska, R.Z.; Jeltsch, A. The DNMT3A R882H mutation does not cause dominant negative effects in purified mixed DNMT3A/R882H complexes. *Sci. Rep.* **2018**, *8*, 13242. [CrossRef] [PubMed]
106. Holz-Schietinger, C.; Matje, D.M.; Reich, N.O. Mutations in DNA methyltransferase (DNMT3A) observed in acute myeloid leukemia patients disrupt processive methylation. *J. Biol. Chem.* **2012**, *287*, 30941–30951. [CrossRef] [PubMed]
107. Russler-Germain, D.A.; Spencer, D.H.; Young, M.A.; Lamprecht, T.L.; Miller, C.A.; Fulton, R.; Meyer, M.R.; Erdmann-Gilmore, P.; Townsend, R.R.; Wilson, R.K.; et al. The R882H DNMT3A mutation associated with AML dominantly inhibits wild-type DNMT3A by blocking its ability to form active tetramers. *Cancer Cell* **2014**, *25*, 442–454. [CrossRef] [PubMed]
108. Otani, J.; Nankumo, T.; Arita, K.; Inamoto, S.; Ariyoshi, M.; Shirakawa, M. Structural basis for recognition of H3K4 methylation status by the DNA methyltransferase 3A ATRX-DNMT3-DNMT3L domain. *EMBO Rep.* **2009**, *10*, 1235–1241. [CrossRef]
109. Zhang, Y.; Jurkowska, R.; Soeroes, S.; Rajavelu, A.; Dhayalan, A.; Bock, I.; Rathert, P.; Brandt, O.; Reinhardt, R.; Fischle, W.; et al. Chromatin methylation activity of Dnmt3a and Dnmt3a/3L is guided by interaction of the ADD domain with the histone H3 tail. *Nucleic Acids Res.* **2010**, *38*, 4246–4253. [CrossRef]

110. Guo, X.; Wang, L.; Li, J.; Ding, Z.; Xiao, J.; Yin, X.; He, S.; Shi, P.; Dong, L.; Li, G.; et al. Structural insight into autoinhibition and histone H3-induced activation of DNMT3A. *Nature* **2014**. [CrossRef]
111. Li, B.Z.; Huang, Z.; Cui, Q.Y.; Song, X.H.; Du, L.; Jeltsch, A.; Chen, P.; Li, G.; Li, E.; Xu, G.L. Histone tails regulate DNA methylation by allosterically activating de novo methyltransferase. *Cell Res.* **2011**, *21*, 1172–1181. [CrossRef] [PubMed]
112. Chen, T.; Ueda, Y.; Xie, S.; Li, E. A novel Dnmt3a isoform produced from an alternative promoter localizes to euchromatin and its expression correlates with active de novo methylation. *J. Biol. Chem.* **2002**, *277*, 38746–38754. [CrossRef] [PubMed]
113. Kotini, A.G.; Mpakali, A.; Agalioti, T. Dnmt3a1 upregulates transcription of distinct genes and targets chromosomal gene clusters for epigenetic silencing in mouse embryonic stem cells. *Mol. Cell. Biol.* **2011**, *31*, 1577–1592. [CrossRef] [PubMed]
114. Suetake, I.; Mishima, Y.; Kimura, H.; Lee, Y.H.; Goto, Y.; Takeshima, H.; Ikegami, T.; Tajima, S. Characterization of DNA-binding activity in the N-terminal domain of the DNA methyltransferase Dnmt3a. *Biochem. J.* **2011**, *437*, 141–148. [CrossRef] [PubMed]
115. Qin, S.; Min, J. Structure and function of the nucleosome-binding PWWP domain. *Trends Biochem. Sci.* **2014**, *39*, 536–547. [CrossRef] [PubMed]
116. Taverna, S.D.; Li, H.; Ruthenburg, A.J.; Allis, C.D.; Patel, D.J. How chromatin-binding modules interpret histone modifications: Lessons from professional pocket pickers. *Nat. Struct. Mol. Biol.* **2007**, *14*, 1025–1040. [CrossRef]
117. Dhayalan, A.; Rajavelu, A.; Rathert, P.; Tamas, R.; Jurkowska, R.Z.; Ragozin, S.; Jeltsch, A. The Dnmt3a PWWP domain reads histone 3 lysine 36 trimethylation and guides DNA methylation. *J. Biol. Chem.* **2010**, *285*, 26114–26120. [CrossRef]
118. Ge, Y.Z.; Pu, M.T.; Gowher, H.; Wu, H.P.; Ding, J.P.; Jeltsch, A.; Xu, G.L. Chromatin targeting of de novo DNA methyltransferases by the PWWP domain. *J. Biol. Chem.* **2004**, *279*, 25447–25454. [CrossRef]
119. Chen, T.; Tsujimoto, N.; Li, E. The PWWP domain of Dnmt3a and Dnmt3b is required for directing DNA methylation to the major satellite repeats at pericentric heterochromatin. *Mol. Cell. Biol.* **2004**, *24*, 9048–9058. [CrossRef]
120. Qiu, C.; Sawada, K.; Zhang, X.; Cheng, X. The PWWP domain of mammalian DNA methyltransferase Dnmt3b defines a new family of DNA-binding folds. *Nat. Struct. Biol.* **2002**, *9*, 217–224. [CrossRef]
121. Rondelet, G.; Dal Maso, T.; Willems, L.; Wouters, J. Structural basis for recognition of histone H3K36me3 nucleosome by human de novo DNA methyltransferases 3A and 3B. *J. Struct. Biol.* **2016**, *194*, 357–367. [CrossRef] [PubMed]
122. Wu, H.; Zeng, H.; Lam, R.; Tempel, W.; Amaya, M.F.; Xu, C.; Dombrovski, L.; Qiu, W.; Wang, Y.; Min, J. Structural and histone binding ability characterizations of human PWWP domains. *PLoS ONE* **2011**, *6*, e18919. [CrossRef] [PubMed]
123. Purdy, M.M.; Holz-Schietinger, C.; Reich, N.O. Identification of a second DNA binding site in human DNA methyltransferase 3A by substrate inhibition and domain deletion. *Arch. Biochem. Biophys.* **2010**, *498*, 13–22. [CrossRef] [PubMed]
124. Neri, F.; Rapelli, S.; Krepelova, A.; Incarnato, D.; Parlato, C.; Basile, G.; Maldotti, M.; Anselmi, F.; Oliviero, S. Intragenic DNA methylation prevents spurious transcription initiation. *Nature* **2017**, *543*, 72–77. [CrossRef] [PubMed]

© 2018 by the authors. Licensee MDPI, Basel, Switzerland. This article is an open access article distributed under the terms and conditions of the Creative Commons Attribution (CC BY) license (http://creativecommons.org/licenses/by/4.0/).

Review

Molecular Processes Connecting DNA Methylation Patterns with DNA Methyltransferases and Histone Modifications in Mammalian Genomes

Albert Jeltsch *, Julian Broche and Pavel Bashtrykov

Institute of Biochemistry and Technical Biochemistry, Department of Biochemistry, University of Stuttgart, 70569 Stuttgart, Germany; Julian.Broche@ibtb.uni-stuttgart.de (J.B.); Pavel.Bashtrykov@ibtb.uni-stuttgart.de (P.B.)
* Correspondence: albert.jeltsch@ibtb.uni-stuttgart.de; Tel.: +49-711-685-64390; Fax: +49-711-685-64392

Received: 25 October 2018; Accepted: 16 November 2018; Published: 21 November 2018

Abstract: DNA methylation is an essential part of the epigenome chromatin modification network, which also comprises several covalent histone protein post-translational modifications. All these modifications are highly interconnected, because the writers and erasers of one mark, DNA methyltransferases (DNMTs) and ten eleven translocation enzymes (TETs) in the case of DNA methylation, are directly or indirectly targeted and regulated by other marks. Here, we have collected information about the genomic distribution and variability of DNA methylation in human and mouse DNA in different genomic elements. After summarizing the impact of DNA methylation on genome evolution including CpG depletion, we describe the connection of DNA methylation with several important histone post-translational modifications, including methylation of H3K4, H3K9, H3K27, and H3K36, but also with nucleosome remodeling. Moreover, we present the mechanistic features of mammalian DNA methyltransferases and their associated factors that mediate the crosstalk between DNA methylation and chromatin modifications. Finally, we describe recent advances regarding the methylation of non-CpG sites, methylation of adenine residues in human cells and methylation of mitochondrial DNA. At several places, we highlight controversial findings or open questions demanding future experimental work.

Keywords: DNA methylation; DNA methyltransferase; histone modification; molecular epigenetics

---

1. Introduction

In mammals, cytosine residues are methylated at levels between 3.5 and 4.5% in adult tissues depending on the cell type; lower levels are observed in embryonic cell lines and rapidly-dividing cells [1,2]. DNA methylation is a major chromatin regulator and an important part of the epigenome network essential for the development of mammals, which functions in concert with other epigenome modifications, most prominently histone tail modifications [3,4]. Aberrant DNA methylation has several connections to diseases including cancer [5,6], and DNA methylation-changing compounds are in development and clinical use for cancer treatment [7,8].

DNA methylation mainly occurs at palindromic CpG sites (28 million sites in the case of the diploid human genome), which are methylated to 70–80%, but cytosines in non-CpG sites are methylated, as well (see below). At CpG sites, the methylation information is present in both DNA strands, meaning that after DNA replication, it can be recovered by a maintenance DNA methyltransferase with high preference for hemimethylated CpG sites, as proposed in the original maintenance DNA methylation model [9] (Figure 1A). Here, we describe DNA methylation patterns in human and mouse DNA in the context of their evolution and compiled information on their correlation with important histone post-translational modifications. Moreover, we describe mechanistic features of mammalian DNA methyltransferases

(DNMTs) that contribute to the crosstalk between DNA methylation and chromatin modifications. Finally, we describe recent advances regarding the methylation of non-CpG sites, methylation of adenine residues in human cells, and methylation of mitochondrial DNA. For more detailed reviews on DNA methylation patterns and functions of DNA methylation, refer to [9–13], and for reviews describing the enzymology of DNMTs to [14–16]. This review will not focus on DNA methylation in other species like plants, fungi, or arthropods, where many (though not all) of the basic processes are conserved, but additional phenomena are observed. We will also not present the details of DNA methylation recognition and processes involved in DNA demethylation. In these fields, the reader is referred to excellent alternative reviews published recently [17–19].

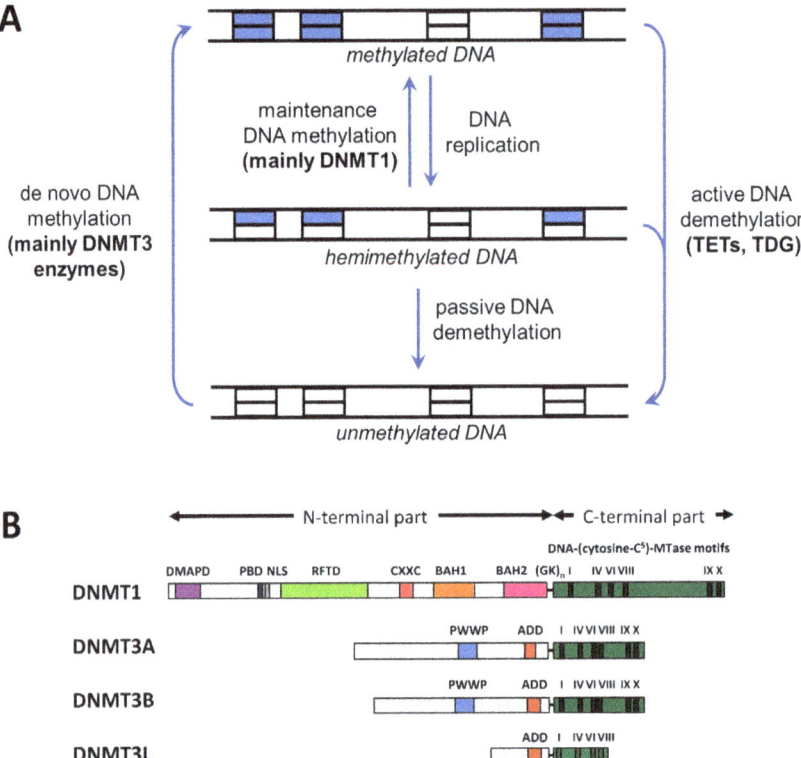

Figure 1. Cycle of DNA methylation and domain structure of DNMTs. (**A**) Cycle of DNA methylation in human cells (adapted from [9]). DNA methylation patterns are generated by de novo methyltransferases and kept through DNA replication by maintenance methylation. DNA methylation can be lost through passive or active demethylation (abbreviations: TET, ten eleven translocation enzyme; TDG, thymine-DNA glycosylase). (**B**) Domain structure of the mammalian DNMTs DNMT1, DNMT3A, and DNMT3B. DNMT3L is a catalytically-inactive member of the DNMT3 family, which has regulatory roles [15]. The human DNMT1, DNMT3A, DNMT3B, and DNMT3L proteins consist of 1616, 912, 853, and 387 amino acid residues, respectively. Abbreviations used: DMAPD, DNA methyltransferase-associated protein 1 interacting domain; PBD, PCNA binding domain; NLS, nuclear localization signal; RFTD, replication foci targeting domain; CXXC, CXXC domain; BAH1 and BAH2, bromo-adjacent homology domains 1 and 2; GK_n, glycine lysine repeats; PWWP, PWWP domain; ADD, ATRX-DNMT3-DNMT3L domain (reprinted from [15] with permission).

DNA methylation is introduced by a family of enzymes called DNA methyltransferases, which all use S-adenosyl-L-methionine as the methyl group donor (reviews: [14–16,20]. In mammals, three active DNMTs are present. DNMT1 is a maintenance methyltransferase with high preference for hemimethylated CpG sites [14,15]. In contrast, the DNMT3A and DNMT3B enzymes do not show preference for hemimethylated target sites, and they are involved in the de novo generation of DNA methylation patterns during germ cell development and the early embryonic phase. All mammalian DNMTs contain a C-terminal catalytic domain, which has structural and sequence homology to prokaryotic DNA-(cytosine C5)-methyltransferases and a larger N-terminal part with different domains involved in targeting and regulation (Figure 1B). DNA demethylation is initiated by the action of the TET family dioxygenases, which catalyze the oxidation of methylcytosine [19].

Despite its overarching elegance, recent data show that the maintenance DNA methylation model cannot fully describe many data, and in fact, all enzymes (DNMT1, DNMT3, and TET enzymes) have roles in DNA methylation after replication and in the de novo generation of DNA methylation and its removal [9]. Therefore, the level of DNA methylation at each cytosine is described by a dynamic equilibrium between gain and loss of methylation [9]. The dynamic nature of DNA methylation patterns has recently been illustrated by showing that the combined knock-out of all TET enzymes leads to hypermethylation of bivalent promoters in human embryonic stem cells (ESC), which was dependent on DNMT3B binding to these sites [21]. Mathematical models have been developed to describe global changes of DNA methylation depending on the expression levels of DNMTs and TETs during serum-to-2i transition of ESCs [22] and gametogenesis [23]. For simulation of local site-specific methylation levels, these models have to be expanded, including the local targeting and preferences of DNMTs and TETs, binding of other proteins, and regulation of DNMTs and TETs, which would usher in a new era of quantitative epigenomics system biology.

2. Evolutionary Impact of DNA Methylation

Due the presence of the methyl group, 5-methylcytosine (5mC) is more prone to deamination, resulting in thymine-guanine (TG) mismatches occurring in a CpG sequence context. This lesion cannot be repaired via the canonical uracil DNA deglycosylase pathway, which otherwise repairs deamination of cytosine, but it requires specific DNA repair enzymes including MBD4 and thymine DNA glycosylase (TGD) in human cells (review: [24]). However, in spite of the presence of these specialized TG mismatch base excision repair systems, the reversal of 5mC deamination is incomplete, and 5mC is mutagenic, which led to a depletion of the genome from methylated CpG sites over evolutionary times. This is illustrated by the fact that the CpG dinucleotide is currently about 4–5-times less abundant than expected on the basis of the single nucleotide frequencies in human DNA (Figure 2), while GpC sites, which are not methylated, are observed roughly at expected frequencies. However, this CpG depletion did not occur (or it is less pronounced) at certain regions, so-called CpG islands (CGI). Typical definitions for CGIs are an average GC frequency of $\geq 50\%$ and CpG observed/expected ratio of ≥ 0.6 in regions of ≥ 400–500 bps [25]. CGIs occur in the promoters of 70% of all genes, and typically, they are not methylated, explaining the local lack of CpG depletion over evolutionary times and the overall reduced depletion of CpG sites in promoters (see below for more details on the methylation of CGIs). Interestingly, the strong selection pressure on exons apparently has drastically diminished the CpG depletion in these genetic elements (Figure 2).

Recently, another unexpected evolutionary impact of DNA methylation and DNMTs has been discovered by showing that different DNMTs (bacterial M.SssI and mouse DNMT3A catalytic domain) also generate low levels of 3 mC [26]. This modified base represents a DNA damage, which is mutagenic and results in a strong replication block. However, this lesion can be directly repaired by ALKB2 family enzymes in an oxidative process [27,28]. Interestingly, it has been found that ALKB2 enzymes are evolutionarily connected with active DNMTs in many species, suggesting their functional relationship [26].

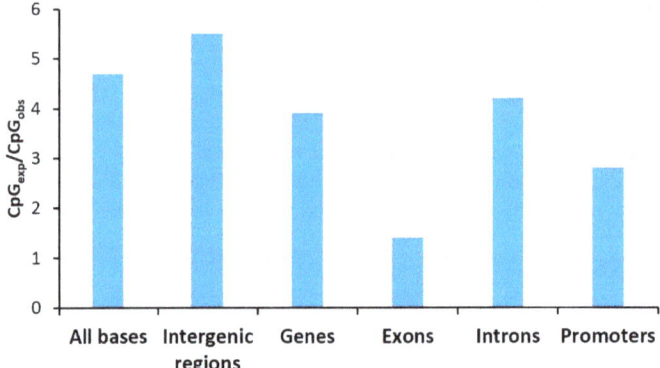

Figure 2. Depletion of CpG dinucleotides in defined genomic elements tabulated for human chromosome 1 [29]. Abbreviations used: CpG_{exp}, expected number of CpG sites (based on the nucleotide composition); CpG_{obs}, observed number of CpG sites.

3. Genomic Distribution and Variability of DNA Methylation

Early studies have demonstrated that the DNA methylation in human DNA shows a biphasic distribution in which CpG sites are either unmethylated or fully methylated [30,31]. Relatively few sites have intermediate methylation levels, which would point towards a heterogeneous methylation state of the corresponding CpG site in the sample. This heterogeneity can arise from cellular heterogeneity, allelic heterogeneity, cell cycle-dependent heterogeneity or fluctuating levels of methylation at one allele. A recent study showed that heterogeneous methylation largely reflects asynchronous proliferation in normal cells, while cancer cells showed more replication-independent heterogeneity [32]. Numerous datasets revealed that DNA methylation is also unequally distributed among genomic elements (Figure 3), and it is correlated with other chromatin marks, which will be described in the following sections in more detail.

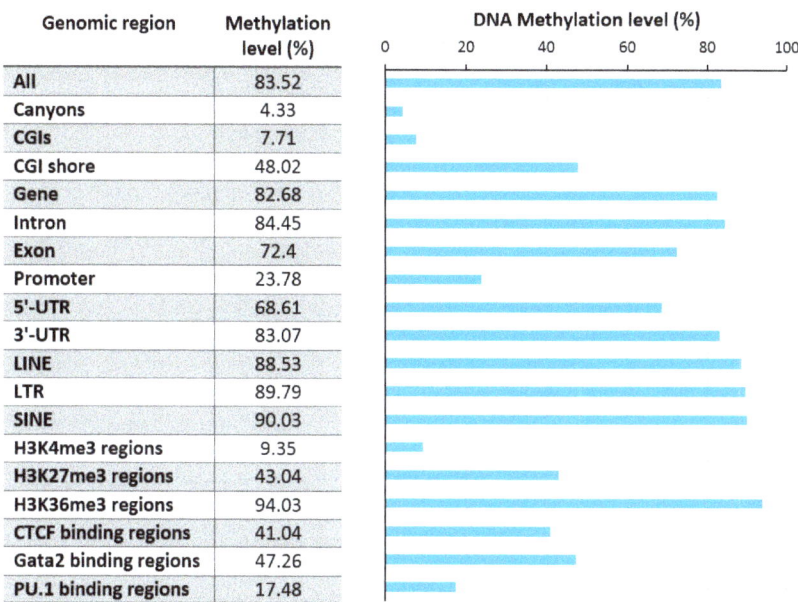

Figure 3. Exemplary DNA methylation levels in mouse hematopoietic stem cells in various genome regions (data taken from [11]). CGI: CpG islands, UTR: untranscribed region, SINE: Short Interspersed Nuclear Elements, LINE: Long Interspersed Nuclear Elements, LTR: Long Terminal Repeat

3.1. Promoter Methylation

There are about 30,000 CGIs in the human genome, and about 70% of all gene promoters are connected to a CGI. CGIs correspond to only 0.67% of the overall genome, but they nevertheless contain approximately 7% of all CpG sites. Methylation of promoter CGIs in normal cells is usually restricted to stable silencing as in X-chromosome inactivation or imprinted genes, but aberrant CGI methylation is observed in cancer cells. However, non-promoter CGIs can become methylated in human tissues in a tissue-specific manner [33]. Moreover, in cell lines and most cancers, hundreds of CGIs are hypermethylated, and in some cases, a so-called CGI methylator phenotype (CIMP) is observed with thousands of CGIs being hypermethylated [34].

Promoters can be differentiated into high CpG (HCG), intermediate CpG (ICG), and low CpG (LCG) promoters, and early genome-wide DNA methylation analyses showed a clear anticorrelation of CpG density and DNA methylation [30,31,35]. Therefore, HCG and LCG usually show low and high DNA methylation, while ICGs show the most variability. These include regions flanking CGIs, so-called CGI shores, which have moderately-elevated CpG frequencies. An analysis of DNA methylation data of 30 human cell and tissue types revealed that methylation of about 20% of the CpG sites are dynamically regulated [36], but this number is likely to increase further as more datasets are incorporated. Most variances in DNA methylation were indeed observed in ICGs and CGI shores.

As mentioned above, CGIs are generally protected from DNA methylation. Different mechanisms appear to be involved in this process: (1) It has been shown that this protection is sequence dependent, suggesting that bound transcription factors (TFs) prevent methylation [37]. Mechanistically, TFs can physically preclude access of DNMTs [38]. Moreover, bound TFs can lead to the deposition of activating marks, particularly H3K4me2/3, which prevents binding of DNMT3 enzymes (see below). The cell type-specific binding of TFs can also explain the strong overrepresentation of regulatory elements in regions that show differential methylation during development [36]. In agreement with this general model, it has been shown that allele-specific methylation often occurs at regulatory sites where single-nucleotide polymorphisms between alleles cause differential binding of TFs [39,40]. (2) CGIs can be protected from methylation by association with TET enzymes. Studies in murine ESCs [41], as well as human embryonic kidney cells (HEK293T) [42] have shown that TET1 strongly binds to CpG-rich DNA associated with high-CpG-density gene promoters and exons, and its density is positively correlated with H3K4me3 in promoter regions. Furthermore, TET2 and TET3 were shown to bind to CpG islands and promoter regions. (3) CXXC domain-containing readers of unmethylated DNA preferentially bind at CGIs, and by this, they recruit other chromatin factors to these regions, which helps to keep them unmethylated [43]. These domains are found in numerous chromatin factors, and they appear to be involved in their targeting to unmethylated CGIs [44], including the KDM2A and KDM2B H3K36-specific lysine demethylases [45], the KMT2A and KMT2B H3K4-specific protein lysine methyltransferases (PKMTs) [46], as well as TET1 and TET3 [47]. This model has recently been further validated experimentally by showing that CXXC domains recruit TET enzymes to unmethylated CpG-rich CGIs, leading to the DNA demethylation and protection of these regions from de novo methylation [21].

Methylated CGIs recruit classical MBD family readers of DNA methylation including MeCP2, MBD1, or MBD2, which establish strong repression of gene expression by containing transcriptional repression domains and forming complexes with other silencing factors including lysine deacetylases, H3K9 methyltransferases (SUV39H1) and chromatin remodelers (review: [18]). In addition, the SETDB1 H3K9 methyltransferase, which has a prominent role in the silencing of repeats and retrotransposons, also contains an MBD domain. However, it has become clear that MBD proteins are not exclusively involved in gene silencing. For example, MeCP2 can function as a gene activator and repressor, as shown by gene expression studies in brain regions where most target genes were found to be upregulated by MeCP2 [48–50]. In agreement with these findings, MeCP2 was shown to interact with DNMT3A and function as an inhibitor or stimulator of DNMT3A activity, depending on the chromatin context [51].

3.2. Enhancer Methylation and Influence of DNA Methylation on TF Binding

Enhancers functionally resemble promoters by containing binding sites for TFs, but they show depletion of CpG sites roughly corresponding to the average genome. Recently, tissue-specific DNA methylation changes have been detected mainly in enhancers during early development and postnatally, which were triggered by de novo methylation and demethylation [52]. Methylation of TF binding sites in CpG-poor promoters or in enhancers can have variable effects on gene expression, depending on its influence on DNA binding of the TF (repelling or enhancing) and the specific role of the TF (activating or repressing). DNA methylation can prevent the binding of several TFs [53,54]. For example, the CTCF protein, composed of a linear array of 11 zinc finger domains, binds to DNA in a methylation-dependent manner, but methylation prevents binding only at some binding sites [55]. This was recently explained by structural data showing that DNA methylation only affects binding if it occurs at one particular position within the consensus binding site [56]. However, DNA methylation can also promote DNA binding of TFs [53,54]. For example, there is a group of C2H2 zinc finger proteins that bind methylated DNA in a sequence-dependent manner [57,58]. One of them is Kaiso, which interacts with the N-CoR repression complex. Another member of this group is ZFP57, a KRAB zinc finger, which recruits the KAP1 corepressor to methylated imprinting control regions. At the cellular level, the modulation of the DNA binding of chromatin-organizing proteins like CTCF and cohesins by DNA methylation has been connected to altered genome structure and cancer [59]. So-called pioneering TFs were shown to bind to heterochromatic and methylated enhancers, leading to their activation, which is accompanied by a slow loss of DNA methylation [60]. These processes are essential steps in the differentiation of cell lineages. Similarly, KLF4 binding to methylated target sites was shown to mediate gene activation [61].

In agreement with an activating role of DNA methylation at some enhancers, it has been shown that H3K27 acetylation and DNA methylation can co-exist at enhancers. These bivalent enhancers were shown to lose acetylation after removal of DNA methylation, implying that DNA methylation was needed for the maintenance of the acetylation mark [62]. Recently, the binding of DNMT3A and DNMT3B to enhancers was studied, showing that both proteins associate with most active enhancers in epidermal stem cells [63]. This binding was dependent on H3K36me3, suggesting that it is mediated by the PWWP domain binding to this mark [15,16]. Interestingly, both DNMT3 proteins differ in their effects on enhancer DNA modification. DNMT3B was shown to be involved in enhancer body methylation, while DNMT3A was shown to cooperate with TET2, promoting enhancer DNA hydroxymethylation. Interestingly, both DNMT3A and DNMT3B are required for enhancer activity and enhancer RNA production, illustrating the dual regulatory potential of DNA methylation for gene repression and activation.

3.3. Repeat Methylation

DNA methylation levels of repeats are generally high, in line with the classical function of DNA methylation to repress the transcriptional activity of repeats and thereby protect genome integrity. There are four classes of highly abundant repeat elements in the human genome: Short Interspersed Nuclear Elements (SINEs), Long Interspersed Nuclear Elements (LINEs), Long Terminal Repeats (LTRs), and DNA transposons (Table 1). All of them show depletion of CpG sites. DNA methylation at repeats varies through development, and ES cells often show reduced repeat methylation levels. In the case of SINE elements, it has been observed that their methylation in ES cells reflects the effect of CpG density on DNA methylation levels in the general genome. CpG-rich SINE elements, which are relatively rare, tend to show low methylation, while CpG-poor elements, which are more abundant, show high methylation [30]. Recently, the repressive role of DNA methylation on repeats has been experimentally documented by showing that treatment of cells with DNMT inhibitors leads to the derepression of LTRs [64,65].

Table 1. Properties and DNA methylation levels of different repeat types in various cell types. Data taken from [66]. The ranges of methylation levels represent the 25th and 75th percentiles. H1 is a human ES cell line; IMR90 are human fetal fibroblasts. Mouse hematopoietic stem cell (mHSC) data were taken from [11].

Repeat Type	Total Number	Mean Length (bps)	Mean GC Content	Mean CpG$_{exp/obs}$	DNA Methylation
SINE	1,426,563	244	0.50	2.98	H1: 81.4–90.2 IMR90: 62.8–90.9 mHSC: 90.0
LINE	947,779	578	0.38	4.01	H1: 82.6–90.7 IMR90: 41.9–88.4 mHSC: 88.53
LTR	530,763	443	0.44	4.84	H1: 81.4–90.7 IMR90: 37.2–83.7 mHSC: 89.79
DNA Transposon	273,586	272	0.40	3.63	H1: 83.7–91.9 IMR90: 45.4–90.7

Targeting of repetitive DNA in part depends on SETDB1 (also known as ESET or KMT1E), a histone H3K9 lysine methyltransferase that generates H3K9me3 in euchromatic regions [67]. It forms a complex with KAP-1 (also called TRIM28) that can further interact with KRAB zinc finger proteins, which are involved in the recognition and silencing of repeats and transposons [68,69]. Moreover, it associates with additional silencing factors including DNMT3A [70]. The KAP-1/SETDB1 complex is required for silencing of LTR retroviruses, and it has also been connected to H3K9 methylation of LINE elements (see the references provided in [67]). Targeting of repetitive DNA in the mammalian germline is also dependent on the piRNA pathway [71], but the details of this process are not known.

3.4. DNA Methylation Canyons

Genome-wide DNA methylation analyses unexpectedly revealed the existence of a new class of large hypomethylated regions, which were called canyons or DNA methylation valleys [72,73].

These regions typically span up to 1 MB and contain several CGIs interrupted by DNA with lower CpG content. DNA methylation canyons are conserved among cell types and species and uniquely enriched for TF binding sites and developmental regulatory genes. Canyon borders were shown to be marked by 5-hydroxymethyl cytosine (5hmC), and they become eroded in the absence of DNMT3A [73], suggesting that the edges of these canyons represent regions where active DNA methylation, hydroxymethylation, and demethylation are in a dynamic steady-state.

4. Relation of DNA Methylation and Chromatin Marks

4.1. H3K4me3

Early genome-wide DNA methylation analyses revealed one of the most striking features of DNA methylation patterns, that DNA methylation is strongly anticorrelated with H3K4me2/3 [30,31], a finding that has been reproduced in several follow-up studies (Figures 3 and 4). H3K4me3 marks active promoters with high occupancy of RNA polymerase II, also showing elevated levels of H3K79me3 (Figure 4). Low DNA methylation of these regions is in agreement with the fact that the ATRX-DNMT3-DNMT3L (ADD) domains of DNMT3A, DNMT3B, and DNMT3L cannot bind to the H3 tail di- or tri-methylated at H3K4 [15,16]. Moreover, the DNMT3 enzymes require binding of the K4 unmodified tail to their ADD domain to activate the catalytic center, such that DNMT3 enzymes aberrantly bound at H3K4me2/3 regions would remain catalytically inactive.

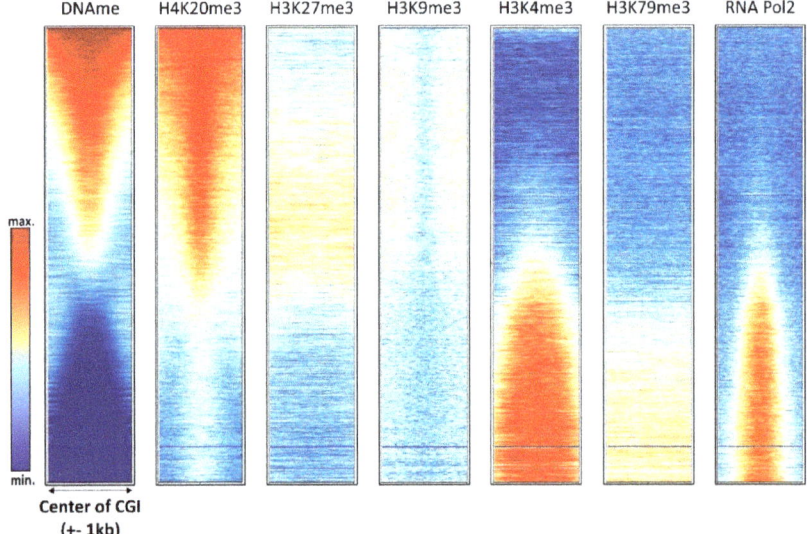

Figure 4. Compilation of DNA methylation density with the density of other chromatin marks in annotated CGIs in HEK293 cells (see Supplementary Text 1 for the data sources).

By this mechanism, H3K4 methylation in active promoters protects these regions from DNA methylation. Mechanistically, the ADD binding of DNMT3 proteins is also disrupted for example by acetylation of K4 or phosphorylation of T3 and T6 [74]. The functional role of the readout of these chromatin marks has been studied experimentally by designing of DNMT3A variants with mutated ADD domains that were no longer sensitive towards K4 methylation or T6 phosphorylation [75]. Expression of these DNMT3A mutants in cells perturbed the differentiation program of ESC, and it led to chromosomal instability. The prominent role of the ADD domain in the targeting of DNMT3 enzymes has recently also been confirmed at enhancers of pluripotency genes, where it was shown that LSD1-dependent demethylation was necessary for DNMTA binding, which led to enhancer methylation [76].

4.2. H3K36me3

H3K36me3 accumulates in the bodies of expressed genes where it is introduced by the SETD2 PKMT, which is recruited by the RNAPII phosphorylated at Ser2 and Ser5 in the C-terminal tail [77,78]. DNA methylation in gene bodies is also particularly high [79], which is in agreement with the binding of the DNMT3A and DNMT3B PWWP domains to the H3K36 methylation mark [15,16]. Within gene bodies, H3K36me3 and DNA methylation are correlated with gene expression, while H3K27me3 and H3K9me3 are anticorrelated with expression (Figures 3 and 5). Using a DNMT3B knock-out mouse ES cell line, it was shown that intragenic DNA methylation is deposited by DNMT3B [80], while the potential role of DNMT3A in this process has remained unclear. Gene body DNA methylation by DNMT3B was shown to be dependent on SETD2-deposited H3K36me3 as expected from the PWWP domain binding this mark [81]. Intragenic DNA methylation has several functions including the regulation of alternative promoters [82] and alternative splicing [83], as well as prevention of intragenic transcription initiation [81], which are in line with the general functions of gene body H3K36 methylation [77]. It is currently unclear if H3K36me3 binding only has a recruiting function for DNMT3 enzymes or if it also regulates their activity. Strikingly, some studies provided evidence that the gene body methylation even has a direct stimulatory role on gene expression by unknown mechanisms [84,85]. Another interesting question is the slightly lower DNA methylation in exons when compared to introns that so far has escaped mechanistic and functional explanation. One possible explanation for this observation could be that exons are more CpG-rich than introns.

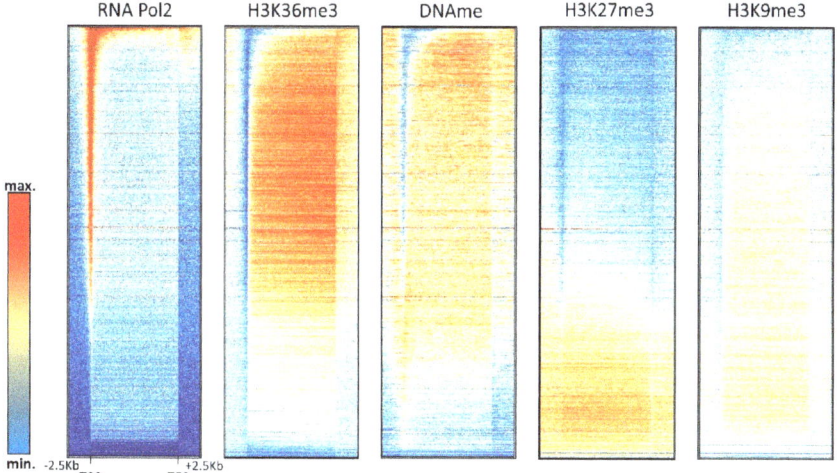

Figure 5. Compilation of DNA methylation density and the density of other chromatin marks in gene bodies in HEK293 cells (see Supplementary Text 1 for the data sources).

4.3. H3K27me3

The relationship of DNA methylation with the Polycomb H3K27me3 mark is complex and ambivalent, because different studies have provided evidence that both marks can either act together or antagonistically. The complex interplay of DNA methylation and H3K27me3 at CGIs is illustrated in Figure 4 showing that highly methylated CGIs in HEK293 cells are depleted from H3K27me3, but CGIs with medium methylation levels show a slight enrichment of H3K27me3. In gene bodies, H3K27me3 seems to be anticorrelated with DNA methylation (Figure 5).

Early studies showed an interaction of DNMTs and PRC2 (the PKMT complex that generates H3K27me3) [86] and demonstrated that H3K27me3 marked genes are targets for aberrant DNA methylation in cancer cells and cell lines [31,87,88]. Later, genome-wide studies in ES cells with knock-out of all DNMTs that are lacking DNA methylation revealed a reduction in localized H3K27me3 peaks, in agreement with the model of a synergistic function of DNA methylation and H3K27me3 [89]. A synergistic function of DNA methylation and H3K27me3 is also supported by the fact that both are enriched at the inactive X-chromosome, where they lead to its transcriptional silencing (review: [90]), and by the recent report that PRC2 preferentially binds CG-rich and CpG methylated DNA in vitro [91].

On the other side, a combined chromatin immunoprecipitation (ChIP)-bisulfite study revealed an antagonism of both marks at CpG islands [89]. In line with this result, other studies found that PRC2 binds to unmethylated, but not to methylated genomic regions, and disruption of DNA methylation leads to the appearance of H3K27me3 at the previously methylated CGIs (review: [92]). The global antagonism of these modifications was also confirmed by showing that DNA methylation valleys contain broad regions of H3K27me3 [72]. It was found that binding of the Polycomb protein, which is part of the PRC1 complex, indeed promotes the hypomethylation of DNA methylation valleys, likely by the regulation of TET enzymes [93].

4.4. H3K9me3

Historically, H3K9me3 is tightly connected with DNA methylation, because the *Neurospora crassa* Dim5 enzyme, one of the first-discovered H3K9-specific PKMTs [94], was found in a screen for mutations with reduced DNA methylation. In mammals, both marks are known to be enriched in heterochromatin, but their connection is not yet fully understood. Knock-out of SUV39H1 and SUV39H2 in ESCs has been shown to cause reduced DNA methylation at major satellite repeats, but not at other different repeat elements [95]. Early studies have shown targeting of DNMT3B to heterochromatic sites by HP1 alpha, an H3K9me3 reader [96]. However, the lack of reduction of CpA methylation (which can only be deposited by DNMT3 enzymes) in SUV39 double-knock-out cells [95] suggested that the connection of H3K9me3 and DNA methylation is rather mediated by DNMT1. One candidate for this function is UHRF1 [97], which is an essential factor for DNA methylation in mammals [15,16]. It was discovered in 2007 that UHFR1 co-localizes with DNMT1 and PCNA at replicating heterochromatic regions during mid- to late S-phase, and the association of DNMT1 with chromatin was lost in UHFR1 knock-out (KO) cells [98,99]. UHFR1 KO is embryonically lethal in mice, and UHRF1-deficient embryos showed strongly reduced levels of genome-wide DNA methylation, indicating that UHRF1 has an essential role in the maintenance of DNA methylation. These observations led to a model that UHFR1 recruits DNMT1 to replicated hemimethylated DNA to facilitate its efficient re-methylation. Since this impressive discovery, the structural, mechanistic, and functional details of the DNMT1-UHRF1 interaction have been a subject of very intense investigation. Like DNMT1, UHRF1 is a large multidomain protein [15]. UHRF1 stimulates the catalytic activity of DNMT1 by an interaction with the DNMT1 RFT domain, which opens the auto-inhibited conformation [100,101].

UHRF1 binds to hemimethylated DNA with its SET- and RING-associated (SRA) domain, and its tandem Tudor domain (TTD) and plant homeodomain (PHD) bind H3-tails containing H3K9me3 and unmodified H3R2 in a cooperative reaction [102,103]. H3K9me3 binding of UHRF1 was required for the localization of UHRF1 to heterochromatin and for maintenance of DNA methylation, since a mutation in TTD, which prevents binding of UHRF1 to H3K9me3, abolished both functions [97,104]. Similarly, disruption of H3R2 binding by the PHD domain abolished DNA methylation by DNMT1 in cells [83]. The reduction of H3K9me2 and UHRF1 observed during global demethylation in the serum-to-2i transition in ESCs was also suggestive of a connection of both processes [22]. However, recently, knock-in of a UHRF1 gene with the mutated H3K9me3 binding site into *Uhfr1* deletion cells led to an almost complete recovery of DNA methylation, suggesting that the H3K9me3 binding site of UHRF1 alone is not essential for DNA methylation [105]. UHRF1 binding also mediates the crosstalk of DNA methylation with asymmetrically-dimethylated H3R2, which is introduced by PRMT6. PRMT6 functions as a negative regulator of DNA methylation, because the H3R2 methylation interferes with UHRF1 binding [106]. By this mechanism, overexpression of PRMT6 in cancer cells could be connected to the global DNA hypomethylation often observed in these cells.

Interestingly, elevated DNA methylation is apparently not accompanied by increased H3K9me3 in promoter CGIs; however, promoter CGI methylation is strongly correlated with H4K20me3 (Figure 4). H4K20me3 is another heterochromatic chromatin modification, deposited by the SUV420H1 and SUV420H2 PKMTs [107]. Its role at promoters and the molecular mechanism of its connection with DNA methylation at these sites has not yet been well investigated. In gene bodies, DNA methylation and H3K9me3 are anticorrelated (Figure 5).

4.5. Chromatin Remodeling and DNA Methylation

Historically, chromatin remodeling has been very tightly connected to DNA methylation, because in an *Arabidopsis thaliana* screen for mutants with lost DNA methylation, the first discovered mutation did not affect a DNA methyltransferase, but the *Ddm1* gene, a putative chromatin remodeler. The mammalian Ddm1 homolog is called HELLS (previously also LSH), and it was shown to be essential for DNA methylation [108,109]. HELLS is an SNF2 ATPase protein and putative chromatin remodeler, and its ATPase activity was indeed shown to be necessary for chromatin binding and stimulation of DNA methylation [110,111]. The close connection of DNA methylation and chromatin remodeling can be understood on the basis of the structures of DNMT1 [112] and DNMT3A [113] in complex with DNA, because both complexes show that DNMT binding would not be possible to nucleosomal DNA. This conclusion is in agreement with experimental data showing that DNA bound to nucleosomes is not efficiently methylated by DNMTs [114–117]. Along the same lines, it was demonstrated that nucleosomal DNA can only be methylated by DNMT1 in the presence of chromatin remodelers [118].

Immunodeficiency-centromeric instability-facial anomalies syndrome (ICF) is a hereditary disease characterized by reduced DNA methylation of pericentromeric heterochromatic repeats mainly in chromosomes 1, 9, and 16, which was initially connected to loss or reduction of function mutations in the DNMT3B DNA methyltransferase [119,120]. Recently, it was shown that in addition to DNMT3B mutations, also mutations in HELLS and the ZBTB24 and CDCA7 Zinc finger proteins can cause ICF [121,122]. ZBTB24 has been shown to promote CDCA7 transcription [123], and the complex of HELLS and CDCA7 recently has been shown to have chromatin remodeling activity [124]. These data suggest that the remodeling activity of HELLS/CDCA7 indeed could be necessary for DNMT3B methylation of the pericentromeric heterochromatic repeats.

5. Additional DNA Methylation Events

5.1. Non-CpG Methylation: Enzymes, Patterns, and Role

Despite clear evidence for the existence of non-CpG methylation in human DNA for a long time, its presence has been documented only during recent years. The reason for this is that it generally occurs at low levels in mammals, which makes it technically difficult to discriminate true non-CpG methylation from incomplete conversion, a regular artifact in bisulfite DNA methylation analysis. Biochemical studies showed that non-CpG methylation can be introduced by DNMT3 enzymes, which have a relaxed specificity [14]. However, DNMT1, the most active and important DNMT in mammals, shows an exquisite specificity and is unable to methylate non-CpG sites [125]. Therefore, while CpG methylation is efficiently duplicated after each cell division, this is not true for non-CpG methylation which must be generated de novo after each DNA replication. Interestingly, this is also true for CHG sites (where H stands for A, T or C), although they are palindromic and contain the methylation information in both DNA strands. In spite of this, DNA methylation at these sites is not "directly maintained", because in mammals there is no DNMT with specificity for hemimethylated CHG sites. Methylation analyses in DNMT KO cell lines clearly demonstrated that DNMT3 enzymes are indeed responsible for non-CpG methylation [95]. Consequently, non-CpG methylation is mainly found in tissues with high expression of DNMT3 enzymes like embryonic tissues and neurons, where the lack of cell division also favors the stable presence of non-CpG methylation. Non-CpG methylation is observed at levels >2% of sites in neurons and at >1% in frontal cortex and embryonic cells [126]. In the genome, it is enriched in gene bodies and transposons. Interestingly, non-CpG methylation mainly occurs in a CAG context in ES cells and in early development, while in differentiated neurons, mainly CAC methylation is observed [126]. Studies in mouse DNMT KO ES cells after re-expression of DNMT3A or DNMT3B revealed that DNMT3A preferred introduction of CAC methylation, while DNMT3B was more active in CAG methylation [127], which is in agreement with the general observation that DNMT3B has a role in early development and that DNMT3A is highly expressed in brain tissues.

In the brain, CpG and non-CpG (mainly CA) methylation are observed in the gene bodies of long MeCP2 repressed genes [128]. Disruption of CA methylation by conditional DNMT3A knock-out revealed that CA methylation is critical for binding of MeCP2 and repression of long genes [129]. In another study, binding of MeCP2 to non-CpG methylated DNA was found to regulate gene expression in both directions in adult mouse brain [130]. A recent study has investigated the distribution of non-CpG methylation on the active and inactive X-chromosome in murine frontal cortex cells showing that non-CpG methylation was enriched in the active X-chromosome and one transcriptionally-active region of the inactive X-chromosome [131], similarly as shown for gene body methylation on the active X-chromosome previously [35]. While this distribution is easily explained by the transcription-dependent targeting of DNMT3 enzymes via H3K36me3 binding of the PWWP domains, it raises interesting new questions regarding the role and function of non-CpG methylation in mammals.

5.2. Rare Methylation Events: Controversies about 6mA and Mitochondrial DNA Methylation

In bacteria, two additional types of naturally-methylated nucleobases occur in DNA, namely 6-methyladenine (6mA) and 4-methylcytosine. For decades, it has been discussed whether particularly the 6mA modification also exists in the DNA of high eukaryotes [132]. With the development of more sensitive liquid-chromatography coupled mass spectrometry (LC-MS) detection systems, several papers recently reported the presence of 6mA in higher eukaryotes including *C. elegans* (0.013–0.39% mdA/dA = 130–3900 ppm, depending on growth conditions) [133], *Drosophila* (10–70 ppm) [134], *Xenopus* (0.9 ppm) [135], and mouse ES cells and tissues (6–7 ppm) [123]. In a more recent paper based on antibody detection, mass spectrometry and single-molecule real-time sequencing reported 6mA levels of 510 ppm [136]. Clearly, even the highest estimates of 6mA in human DNA are below 10^6 modified adenine residues per diploid genome, indicating that it is a rare modification. The presence of 6mA in mouse DNA has recently been reexamined using highly sophisticated quantitative LC-MS methods [137]. These authors did not find evidence for 6mA in mouse ES cells and tissues with a detection limit of 0.35 ppm. They also reported 6mA contaminations in commercial enzyme preparations and highlighted the possibility of 6mA contamination via bacterial DNA (from skin or gut) or after ingestion of bacterial DNA and incorporation of 6mA into the DNA through the salvage pathway. More carefully-controlled experiments and cross-lab validations will be necessary to clarify this question.

A similar controversy exists about mitochondrial DNA methylation, where papers presented evidence in favor of its existence (mostly based on bisulfite conversion coupled with sequencing) [138]. Mitochondrial DNA methylation had been connected with a mitochondrial isoform of DNMT1 [139], which is in apparent disagreement with another paper reporting that mitochondrial DNA methylation occurs at GpC (not CpG) sites [140], because DNMT1 is a strict CpG methyltransferase. Different recent bisulfite conversion studies coupled with deep sequencing provided evidence that most mitochondrial DNA methylation signals may be caused by technical artifacts of the bisulfite sequencing technology [141,142]. These controversies illustrate that particularly in the cases of low to very low methylation levels, powerful controls must be included in any analysis, and the biological relevance of potential methylation events must be rigorously assessed.

6. Outlook

DNA methylation is an essential part of the chromatin modification network, also comprising several covalent histone protein post-translational modifications. All these modifications are highly interconnected, because the writers and erasers of one mark (DNMTs and TETs in the case of DNA methylation) are directly or indirectly targeted and regulated by other marks. Our understanding of the properties of this epigenome network are still immature, and many more experiments investigating individual functional connections between the chromatin marks, but also the global effects will be needed for a detailed quantitative description of this network. In the specific case of DNA methylation, we need to understand better how epigenome marks and chromatin structure target and regulate DNMTs and TET enzymes. A better understanding of the functional connections between different epigenome marks will greatly improve our understanding of developmental processes and finally also propel our abilities in epigenome editing, a bioengineering approach aiming at the durable editing of individual chromatin marks like DNA methylation specifically at defined genome loci, which has many promising applications both in basic research and in molecular medicine [143,144].

Supplementary Materials: The following documents are available online at http://www.mdpi.com/2073-4425/9/11/566/s1. Supplemental text 1: Sources of the datasets shown in Figure 4.

Funding: Work in the authors' laboratory has been supported by the DFG (JE 252/6 and JE 252/15), the Baden-Württemberg (BW) Foundation (BWST_NCRNA_007), and the Bundesministerium für Bildung und Forschung BMBF (01GM1513E).

Conflicts of Interest: The authors declare no conflict of interest.

References

1. Globisch, D.; Münzel, M.; Müller, M.; Michalakis, S.; Wagner, M.; Koch, S.; Brückl, T.; Biel, M.; Carell, T. Tissue distribution of 5-hydroxymethylcytosine and search for active demethylation intermediates. *PLoS ONE* **2010**, *5*, e15367. [CrossRef] [PubMed]
2. Lisanti, S.; Omar, W.A.; Tomaszewski, B.; De Prins, S.; Jacobs, G.; Koppen, G.; Mathers, J.C.; Langie, S.A. Comparison of methods for quantification of global DNA methylation in human cells and tissues. *PLoS ONE* **2013**, *8*, e79044. [CrossRef] [PubMed]
3. Allis, C.D.; Jenuwein, T. The molecular hallmarks of epigenetic control. *Nat. Rev. Genet.* **2016**, *17*, 487–500. [CrossRef] [PubMed]
4. Noh, K.M.; Allis, C.D.; Li, H. Reading between the lines: "ADD"-ing histone and DNA methylation marks toward a new epigenetic "Sum". *ACS Chem. Biol.* **2018**, *13*, 1103. [CrossRef] [PubMed]
5. Baylin, S.B.; Jones, P.A. A decade of exploring the cancer epigenome—Biological and translational implications. *Nat. Rev. Cancer* **2011**, *11*, 726–734. [CrossRef] [PubMed]
6. Bergman, Y.; Cedar, H. DNA methylation dynamics in health and disease. *Nat. Struct. Mol. Biol.* **2013**, *20*, 274–281. [CrossRef] [PubMed]
7. Erdmann, A.; Halby, L.; Fahy, J.; Arimondo, P.B. Targeting DNA methylation with small molecules: What's next? *J. Med. Chem.* **2015**, *58*, 2569–2583. [CrossRef] [PubMed]
8. Mirfattah, B.; Herring, J.; Tang, H.; Zhang, K. Probes and targets of DNA methylation and demethylation in drug development. *Curr. Top. Med. Chem.* **2017**, *17*, 1727–1740. [CrossRef] [PubMed]
9. Jeltsch, A.; Jurkowska, R.Z. New concepts in DNA methylation. *Trends Biochem. Sci.* **2014**, *39*, 310–318. [CrossRef] [PubMed]
10. Jones, P.A. Functions of DNA methylation: Islands, start sites, gene bodies and beyond. *Nat. Rev. Genet.* **2012**, *13*, 484–492. [CrossRef] [PubMed]
11. Jeong, M.; Goodell, M.A. New answers to old questions from genome-wide maps of DNA methylation in hematopoietic cells. *Exp. Hematol.* **2014**, *42*, 609–617. [CrossRef] [PubMed]
12. Schubeler, D. Function and information content of DNA methylation. *Nature* **2015**, *517*, 321–326. [CrossRef] [PubMed]

13. Ye, F.; Kong, X.; Zhang, H.; Liu, Y.; Shao, Z.; Jin, J.; Cai, Y.; Zhang, R.; Li, L.; Zhang, Y.W.; et al. Biochemical studies and molecular dynamic simulations reveal the molecular basis of conformational changes in DNA methyltransferase-1. *ACS Chem. Biol.* **2018**, *13*, 772–781. [CrossRef] [PubMed]
14. Jurkowska, R.Z.; Jurkowski, T.P.; Jeltsch, A. Structure and function of mammalian DNA methyltransferases. *Chembiochem* **2011**, *12*, 206–222. [CrossRef] [PubMed]
15. Jeltsch, A.; Jurkowska, R.Z. Allosteric control of mammalian DNA methyltransferases—A new regulatory paradigm. *Nucleic Acids Res.* **2016**, *44*, 8556–8575. [CrossRef] [PubMed]
16. Gowher, H.; Jeltsch, A. Mammalian DNA methyltransferases: New discoveries and open questions. *Biochem. Soc. Trans.* **2018**, *46*, 1191–1202. [CrossRef] [PubMed]
17. Law, J.A.; Jacobsen, S.E. Establishing, maintaining and modifying DNA methylation patterns in plants and animals. *Nat. Rev. Genet.* **2010**, *11*, 204–220. [CrossRef] [PubMed]
18. Du, J.; Johnson, L.M.; Jacobsen, S.E.; Patel, D.J. DNA methylation pathways and their crosstalk with histone methylation. *Nat. Rev. Mol. Cell Biol.* **2015**, *16*, 519–532. [CrossRef] [PubMed]
19. Wu, X.; Zhang, Y. TET-mediated active DNA demethylation: Mechanism, function and beyond. *Nat. Rev. Genet.* **2017**, *18*, 517–534. [CrossRef] [PubMed]
20. Jeltsch, A. Beyond Watson and Crick: DNA methylation and molecular enzymology of DNA methyltransferases. *Chembiochem* **2002**, *3*, 275–293. [CrossRef]
21. Verma, N.; Pan, H.; Doré, L.C.; Shukla, A.; Li, Q.V.; Pelham-Webb, B.; Teijeiro, V.; González, F.; Krivtsov, A.; Chang, C.J.; et al. TET proteins safeguard bivalent promoters from de novo methylation in human embryonic stem cells. *Nat. Genet.* **2018**, *50*, 83–95. [CrossRef] [PubMed]
22. Von Meyenn, F.; Iurlaro, M.; Habibi, E.; Liu, N.Q.; Salehzadeh-Yazdi, A.; Santos, F.; Petrini, E.; Milagre, I.; Yu, M.; Xie, Z.; et al. Impairment of DNA methylation maintenance is the main cause of global demethylation in naive embryonic stem cells. *Mol. Cell* **2016**, *62*, 848–861. [CrossRef] [PubMed]
23. Gahurova, L.; Tomizawa, S.I.; Smallwood, S.A.; Stewart-Morgan, K.R.; Saadeh, H.; Kim, J.; Andrews, S.R.; Chen, T.; Kelsey, G.; et al. Transcription and chromatin determinants of de novo DNA methylation timing in oocytes. *Epigenetics Chromatin* **2017**, *10*, 25. [CrossRef] [PubMed]
24. Bellacosa, A.; Drohat, A.C. Role of base excision repair in maintaining the genetic and epigenetic integrity of CpG sites. *DNA Repair (Amst)* **2015**, *32*, 33–42. [CrossRef] [PubMed]
25. Illingworth, R.S.; Bird, A.P. CpG islands—'A rough guide'. *FEBS Lett.* **2009**, *583*, 1713–1720. [CrossRef] [PubMed]
26. Rosic, S.; Amouroux, R.; Requena, C.E.; Gomes, A.; Emperle, M.; Beltran, T.; Rane, J.K.; Linnett, S.; Selkirk, M.E.; Schiffer, P.H.; et al. Evolutionary analysis indicates that DNA alkylation damage is a byproduct of cytosine DNA methyltransferase activity. *Nat. Genet.* **2018**, *50*, 452–459. [CrossRef] [PubMed]
27. Shen, L.; Song, C.X.; He, C.; Zhang, Y. Mechanism and function of oxidative reversal of DNA and RNA methylation. *Annu. Rev. Biochem.* **2014**, *83*, 585–614. [CrossRef] [PubMed]
28. Fedeles, B.I.; Singh, V.; Delaney, J.C.; Li, D.; Essigmann, J.M. The AlkB Family of Fe(II)/α-ketoglutarate-dependent dioxygenases: Repairing nucleic acid alkylation damage and beyond. *J. Biol. Chem.* **2015**, *290*, 20734–20742. [CrossRef] [PubMed]
29. Wojciechowski, M.; Czapinska, H.; Bochtler, M. CpG underrepresentation and the bacterial CpG-specific DNA methyltransferase M.MpeI. *Proc. Natl. Acad. Sci. USA* **2013**, *110*, 105–110. [CrossRef] [PubMed]
30. Meissner, A.; Mikkelsen, T.S.; Gu, H.; Wernig, M.; Hanna, J.; Sivachenko, A.; Zhang, X.; Bernstein, B.E.; Nusbaum, C.; Jaffe, D.B.; et al. Genome-scale DNA methylation maps of pluripotent and differentiated cells. *Nature* **2008**, *454*, 766–770. [CrossRef] [PubMed]
31. Zhang, Y.; Rohde, C.; Tierling, S.; Jurkowski, T.P.; Bock, C.; Santacruz, D.; Ragozin, S.; Reinhardt, R.; Groth, M.; Walter, J.; et al. DNA methylation analysis of chromosome 21 gene promoters at single base pair and single allele resolution. *PLoS Genet.* **2009**, *5*, e1000438. [CrossRef] [PubMed]
32. Charlton, J.; Downing, T.L.; Smith, Z.D.; Gu, H.; Clement, K.; Pop, R.; Akopian, V.; Klages, S.; Santos, D.P.; Tsankov, A.M.; et al. Global delay in nascent strand DNA methylation. *Nat. Struct. Mol. Biol.* **2018**, *25*, 327–332. [CrossRef] [PubMed]
33. Lokk, K.; Modhukur, V.; Rajashekar, B.; Märtens, K.; Mägi, R.; Kolde, R.; Koltšina, M.; Nilsson, T.K.; Vilo, J.; Salumets, A.; et al. DNA methylome profiling of human tissues identifies global and tissue-specific methylation patterns. *Genome Biol.* **2014**, *15*, r54. [CrossRef] [PubMed]
34. Pfeifer, G.P. Defining driver DNA methylation changes in human cancer. *Int. J. Mol. Sci.* **2018**, *19*, 1166. [CrossRef] [PubMed]

35. Weber, M.; Hellmann, I.; Stadler, M.B.; Ramos, L.; Pääbo, S.; Rebhan, M.; Schübeler, D. Distribution, silencing potential and evolutionary impact of promoter DNA methylation in the human genome. *Nat. Genet.* **2007**, *39*, 457–466. [CrossRef] [PubMed]
36. Ziller, M.J.; Gu, H.; Müller, F.; Donaghey, J.; Tsai, L.T.; Kohlbacher, O.; De Jager, P.L.; Rosen, E.D.; Bennett, D.A.; Bernstein, B.E.; et al. Charting a dynamic DNA methylation landscape of the human genome. *Nature* **2013**, *500*, 477–481. [CrossRef] [PubMed]
37. Long, H.K.; King, H.W.; Patient, R.K.; Odom, D.T.; Klose, R.J. Protection of CpG islands from DNA methylation is DNA-encoded and evolutionarily conserved. *Nucleic Acids Res.* **2016**, *44*, 6693–6706. [CrossRef] [PubMed]
38. Stadler, M.B.; Murr, R.; Burger, L.; Ivanek, R.; Lienert, F.; Schöler, A.; van Nimwegen, E.; Wirbelauer, C.; Oakeley, E.J.; Gaidatzis, D.; et al. DNA-binding factors shape the mouse methylome at distal regulatory regions. *Nature* **2011**, *480*, 490–495. [CrossRef] [PubMed]
39. Onuchic, V.; Lurie, E.; Carrero, I.; Pawliczek, P.; Patel, R.Y.; Rozowsky, J.; Galeev, T.; Huang, Z.; Altshuler, R.C.; Zhang, Z.; et al. Allele-specific epigenome maps reveal sequence-dependent stochastic switching at regulatory loci. *Science* **2018**, *361*. [CrossRef] [PubMed]
40. Zhang, Y.; Rohde, C.; Reinhardt, R.; Voelcker-Rehage, C.; Jeltsch, A. Non-imprinted allele-specific DNA methylation on human autosomes. *Genome Biol.* **2009**, *10*, R138. [CrossRef] [PubMed]
41. Xu, Y.; Wu, F.; Tan, L.; Kong, L.; Xiong, L.; Deng, J.; Barbera, A.J.; Zheng, L.; Zhang, H.; Huang, S.; et al. Genome-wide regulation of 5hmC, 5mC, and gene expression by Tet1 hydroxylase in mouse embryonic stem cells. *Mol. Cell* **2011**, *42*, 451–464. [CrossRef] [PubMed]
42. Pastor, W.A.; Pape, U.J.; Huang, Y.; Henderson, H.R.; Lister, R.; Ko, M.; McLoughlin, E.M.; Brudno, Y.; Mahapatra, S.; Kapranov, P.; et al. Genome-wide mapping of 5-hydroxymethylcytosine in embryonic stem cells. *Nature* **2011**, *473*, 394–397. [CrossRef] [PubMed]
43. Xu, C.; Liu, K.; Lei, M.; Yang, A.; Li, Y.; Hughes, T.R.; Min, J. DNA Sequence recognition of human CXXC domains and their structural determinants. *Structure* **2018**, *26*, 85–95. [CrossRef] [PubMed]
44. Long, H.K.; Blackledge, N.P.; Klose, R.J. ZF-CxxC domain-containing proteins, CpG islands and the chromatin connection. *Biochem. Soc. Trans.* **2013**, *41*, 727–740. [CrossRef] [PubMed]
45. Tanaka, Y.; Umata, T.; Okamoto, K.; Obuse, C.; Tsuneoka, M. CxxC-ZF domain is needed for KDM2A to demethylate histone in rDNA promoter in response to starvation. *Cell Struct. Funct.* **2014**, *39*, 79–92. [CrossRef] [PubMed]
46. Hashimoto, H.; Vertino, P.M.; Cheng, X. Molecular coupling of DNA methylation and histone methylation. *Epigenomics* **2010**, *2*, 657–669. [CrossRef] [PubMed]
47. Melamed, P.; Yosefzon, Y.; David, C.; Tsukerman, A.; Pnueli, L. TET enzymes, variants, and differential effects on function. *Front. Cell Dev. Biol.* **2018**, *6*, 22. [CrossRef] [PubMed]
48. Chahrour, M.; Jung, S.Y.; Shaw, C.; Zhou, X.; Wong, S.T.; Qin, J.; Zoghbi, H.Y. MeCP2, a key contributor to neurological disease, activates and represses transcription. *Science* **2008**, *320*, 1224–1229. [CrossRef] [PubMed]
49. Ben-Shachar, S.; Chahrour, M.; Thaller, C.; Shaw, C.A.; Zoghbi, H.Y. Mouse models of MeCP2 disorders share gene expression changes in the cerebellum and hypothalamus. *Hum. Mol. Genet.* **2009**, *18*, 2431–2442. [CrossRef] [PubMed]
50. Sugino, K.; Hempel, C.M.; Okaty, B.W.; Arnson, H.A.; Kato, S.; Dani, V.S.; Nelson, S.B. Cell-type-specific repression by methyl-CpG-binding protein 2 is biased toward long genes. *J. Neurosci.* **2014**, *34*, 12877–12883. [CrossRef] [PubMed]
51. Rajavelu, A.; Lungu, C.; Emperle, M.; Dukatz, M.; Bröhm, A.; Broche, J.; Hanelt, I.; Parsa, E.; Schiffers, S.; Karnik, R.; et al. Chromatin-dependent allosteric regulation of DNMT3A activity by MeCP2. *Nucleic Acids Res.* **2018**, *46*, 9044–9056. [CrossRef] [PubMed]
52. Reizel, Y.; Sabag, O.; Skversky, Y.; Spiro, A.; Steinberg, B.; Bernstein, D.; Wang, A.; Kieckhaefer, J.; Li, C.; Pikarsky, E.; et al. Postnatal DNA demethylation and its role in tissue maturation. *Nat. Commun.* **2018**, *9*, 2040. [CrossRef] [PubMed]
53. Yin, Y.; Morgunova, E.; Jolma, A.; Kaasinen, E.; Sahu, B.; Khund-Sayeed, S.; Das, P.K.; Kivioja, T.; Dave, K.; Zhong, F.; et al. Impact of cytosine methylation on DNA binding specificities of human transcription factors. *Science* **2017**, *356*. [CrossRef] [PubMed]

54. Kribelbauer, J.F.; Laptenko, O.; Chen, S.; Martini, G.D.; Freed-Pastor, W.A.; Prives, C.; Mann, R.S.; Bussemaker, H.J. Quantitative analysis of the dna methylation sensitivity of transcription factor complexes. *Cell Rep.* **2017**, *19*, 2383–2395. [CrossRef] [PubMed]
55. Maurano, M.T.; Wang, H.; John, S.; Shafer, A.; Canfield, T.; Lee, K.; Stamatoyannopoulos, J.A. Role of DNA methylation in modulating transcription factor occupancy. *Cell Rep.* **2015**, *12*, 1184–1195. [CrossRef] [PubMed]
56. Hashimoto, H.; Wang, D.; Horton, J.R.; Zhang, X.; Corces, V.G.; Cheng, X. Structural basis for the versatile and methylation-dependent binding of CTCF to DNA. *Mol. Cell* **2017**, *66*, 711–720. [CrossRef] [PubMed]
57. Baubec, T.; Schubeler, D. Genomic patterns and context specific interpretation of DNA methylation. *Curr. Opin. Genet. Dev.* **2014**, *25*, 85–92. [CrossRef] [PubMed]
58. Shimbo, T.; Wade, P.A. Proteins that read DNA methylation. *Adv. Exp. Med. Biol.* **2016**, *945*, 303–320. [PubMed]
59. Flavahan, W.A.; Drier, Y.; Liau, B.B.; Gillespie, S.M.; Venteicher, A.S.; Stemmer-Rachamimov, A.O.; Suvà, M.L.; Bernstein, B.E. Insulator dysfunction and oncogene activation in IDH mutant gliomas. *Nature* **2016**, *529*, 110–114. [CrossRef] [PubMed]
60. Mayran, A.; Khetchoumian, K.; Hariri, F.; Pastinen, T.; Gauthier, Y.; Balsalobre, A.; Drouin, J. Pioneer factor Pax7 deploys a stable enhancer repertoire for specification of cell fate. *Nat. Genet.* **2018**, *50*, 259–269. [CrossRef] [PubMed]
61. Wan, J.; Su, Y.; Song, Q.; Tung, B.; Oyinlade, O.; Liu, S.; Ying, M.; Ming, G.L.; Song, H.; Qian, J.; et al. Methylated cis-regulatory elements mediate KLF4-dependent gene transactivation and cell migration. *Elife* **2017**, *6*. [CrossRef] [PubMed]
62. Charlet, J.; Duymich, C.E.; Lay, F.D.; Mundbjerg, K.; Dalsgaard Sørensen, K.; Liang, G.; Jones, P.A. Bivalent regions of cytosine methylation and H3K27 acetylation suggest an active role for DNA methylation at enhancers. *Mol. Cell* **2016**, *62*, 422–431. [CrossRef] [PubMed]
63. Rinaldi, L.; Datta, D.; Serrat, J.; Morey, L.; Solanas, G.; Avgustinova, A.; Blanco, E.; Pons, J.I.; Matallanas, D.; Von Kriegsheim, A.; et al. Dnmt3a and Dnmt3b associate with enhancers to regulate human epidermal stem cell homeostasis. *Cell Stem Cell* **2016**, *19*, 491–501. [CrossRef] [PubMed]
64. Roulois, D.; Loo Yau, H.; Singhania, R.; Wang, Y.; Danesh, A.; Shen, S.Y.; Han, H.; Liang, G.; Jones, P.A.; Pugh, T.J.; et al. DNA-demethylating agents target colorectal cancer cells by inducing viral mimicry by endogenous transcripts. *Cell* **2015**, *162*, 961–973. [CrossRef] [PubMed]
65. Brocks, D.; Schmidt, C.R.; Daskalakis, M.; Jang, H.S.; Shah, N.M.; Li, D.; Li, J.; Zhang, B.; Hou, Y.; Laudato, S.; et al. DNMT and HDAC inhibitors induce cryptic transcription start sites encoded in long terminal repeats. *Nat. Genet.* **2017**, *49*, 1052–1060. [CrossRef] [PubMed]
66. Su, J.; Shao, X.; Liu, H.; Liu, S.; Wu, Q.; Zhang, Y. Genome-wide dynamic changes of DNA methylation of repetitive elements in human embryonic stem cells and fetal fibroblasts. *Genomics* **2012**, *99*, 10–17. [CrossRef] [PubMed]
67. Jurkowska, R.Z.; Qin, S.; Kungulovski, G.; Tempel, W.; Liu, Y.; Bashtrykov, P.; Stiefelmaier, J.; Jurkowski, T.P.; Kudithipudi, S.; Weirich, S.; et al. H3K14ac is linked to methylation of H3K9 by the triple Tudor domain of SETDB1. *Nat. Commun.* **2017**, *8*, 2057. [CrossRef] [PubMed]
68. Lupo, A.; Cesaro, E.; Montano, G.; Zurlo, D.; Izzo, P.; Costanzo, P. KRAB-Zinc finger proteins: A repressor family displaying multiple biological functions. *Curr. Genomics* **2013**, *14*, 268–278. [CrossRef] [PubMed]
69. Fasching, L.; Kapopoulou, A.; Sachdeva, R.; Petri, R.; Jönsson, M.E.; Männe, C.; Turelli, P.; Jern, P.; Cammas, F.; Trono, D.; et al. TRIM28 represses transcription of endogenous retroviruses in neural progenitor cells. *Cell Rep.* **2015**, *10*, 20–28. [CrossRef] [PubMed]
70. Egger, G.; Jeong, S.; Escobar, S.G.; Cortez, C.C.; Li, T.W.; Saito, Y.; Yoo, C.B.; Jones, P.A.; Liang, G. Identification of DNMT1 (DNA methyltransferase 1) hypomorphs in somatic knockouts suggests an essential role for DNMT1 in cell survival. *Proc. Natl. Acad. Sci. USA* **2006**, *103*, 14080–14085. [CrossRef] [PubMed]
71. Siomi, M.C.; Sato, K.; Pezic, D.; Aravin, A.A. PIWI-interacting small RNAs: The vanguard of genome defence. *Nat. Rev. Mol. Cell Biol.* **2011**, *12*, 246–258. [CrossRef] [PubMed]
72. Xie, W.; Schultz, M.D.; Lister, R.; Hou, Z.; Rajagopal, N.; Ray, P.; Whitaker, J.W.; Tian, S.; Hawkins, R.D.; Leung, D.; et al. Epigenomic analysis of multilineage differentiation of human embryonic stem cells. *Cell* **2013**, *153*, 1134–1148. [CrossRef] [PubMed]
73. Jeong, M.; Sun, D.; Luo, M.; Huang, Y.; Challen, G.A.; Rodriguez, B.; Zhang, X.; Chavez, L.; Wang, H.; Hannah, R.; et al. Large conserved domains of low DNA methylation maintained by Dnmt3a. *Nat. Genet.* **2014**, *46*, 17–23. [CrossRef] [PubMed]

74. Zhang, Y.; Jurkowska, R.; Soeroes, S.; Rajavelu, A.; Dhayalan, A.; Bock, I.; Rathert, P.; Brandt, O.; Reinhardt, R.; Fischle, W.; et al. Chromatin methylation activity of Dnmt3a and Dnmt3a/3L is guided by interaction of the ADD domain with the histone H3 tail. *Nucleic Acids Res.* **2010**, *38*, 4246–4253. [CrossRef] [PubMed]
75. Noh, K.M.; Wang, H.; Kim, H.R.; Wenderski, W.; Fang, F.; Li, C.H.; Dewell, S.; Hughes, S.H.; Melnick, A.M.; Patel, D.J.; et al. Engineering of a histone-recognition domain in Dnmt3a alters the epigenetic landscape and phenotypic features of mouse ESCs. *Mol. Cell* **2015**, *59*, 89–103. [CrossRef] [PubMed]
76. Petell, C.J.; Alabdi, L.; He, M.; San Miguel, P.; Rose, R.; Gowher, H. An epigenetic switch regulates de novo DNA methylation at a subset of pluripotency gene enhancers during embryonic stem cell differentiation. *Nucleic Acids Res.* **2016**, *44*, 7605–7617. [CrossRef] [PubMed]
77. Wagner, E.J.; Carpenter, P.B. Understanding the language of Lys36 methylation at histone H3. *Nat. Rev. Mol. Cell Biol.* **2012**, *13*, 115–126. [CrossRef] [PubMed]
78. McDaniel, S.L.; Strahl, B.D. Shaping the cellular landscape with Set2/SETD2 methylation. *Cell. Mol. Life Sci.* **2017**, *74*, 3317–3334. [CrossRef] [PubMed]
79. Ball, M.P.; Li, J.B.; Gao, Y.; Lee, J.H.; LeProust, E.M.; Park, I.H.; Xie, B.; Daley, G.Q.; Church, G.M. Targeted and genome-scale strategies reveal gene-body methylation signatures in human cells. *Nat. Biotechnol.* **2009**, *27*, 361–368. [CrossRef] [PubMed]
80. Baubec, T.; Colombo, D.F.; Wirbelauer, C.; Schmidt, J.; Burger, L.; Krebs, A.R.; Akalin, A.; Schübeler, D. Genomic profiling of DNA methyltransferases reveals a role for DNMT3B in genic methylation. *Nature* **2015**, *520*, 243–247. [CrossRef] [PubMed]
81. Neri, F.; Rapelli, S.; Krepelova, A.; Incarnato, D.; Parlato, C.; Basile, G.; Maldotti, M.; Anselmi, F.; Oliviero, S. Intragenic DNA methylation prevents spurious transcription initiation. *Nature* **2017**, *543*, 72–77. [CrossRef] [PubMed]
82. Maunakea, A.K.; Nagarajan, R.P.; Bilenky, M.; Ballinger, T.J.; D'Souza, C.; Fouse, S.D.; Johnson, B.E.; Hong, C.; Nielsen, C.; Zhao, Y.; et al. Conserved role of intragenic DNA methylation in regulating alternative promoters. *Nature* **2010**, *466*, 253–257. [CrossRef] [PubMed]
83. Qin, W.; Wolf, P.; Liu, N.; Link, S.; Smets, M.; La Mastra, F.; Forné, I.; Pichler, G.; Hörl, D.; Fellinger, K.; et al. DNA methylation requires a DNMT1 ubiquitin interacting motif (UIM) and histone ubiquitination. *Cell Res.* **2015**, *25*, 911–929. [CrossRef] [PubMed]
84. Yang, X.; Han, H.; De Carvalho, D.D.; Lay, F.D.; Jones, P.A.; Liang, G. Gene body methylation can alter gene expression and is a therapeutic target in cancer. *Cancer Cell* **2014**, *26*, 577–590. [CrossRef] [PubMed]
85. Su, J.; Huang, Y.H.; Cui, X.; Wang, X.; Zhang, X.; Lei, Y.; Xu, J.; Lin, X.; Chen, K.; Lv, J.; et al. Homeobox oncogene activation by pan-cancer DNA hypermethylation. *Genome Biol.* **2018**, *19*, 108. [CrossRef] [PubMed]
86. Vire, E.; Brenner, C.; Deplus, R.; Blanchon, L.; Fraga, M.; Didelot, C.; Morey, L.; Van Eynde, A.; Bernard, D.; Vanderwinden, J.M. The Polycomb group protein EZH2 directly controls DNA methylation. *Nature* **2006**, *439*, 871–874. [CrossRef] [PubMed]
87. Widschwendter, M.; Fiegl, H.; Egle, D.; Mueller-Holzner, E.; Spizzo, G.; Marth, C.; Weisenberger, D.J.; Campan, M.; Young, J.; Jacobs, I.; et al. Epigenetic stem cell signature in cancer. *Nat. Genet.* **2007**, *39*, 157–158. [CrossRef] [PubMed]
88. Schlesinger, Y.; Straussman, R.; Keshet, I.; Farkash, S.; Hecht, M.; Zimmerman, J.; Eden, E.; Yakhini, Z.; Ben-Shushan, E.; Reubinoff, B.E.; et al. Polycomb-mediated methylation on Lys27 of histone H3 pre-marks genes for de novo methylation in cancer. *Nat. Genet.* **2007**, *39*, 232–236. [CrossRef] [PubMed]
89. Brinkman, A.B.; Gu, H.; Bartels, S.J.; Zhang, Y.; Matarese, F.; Simmer, F.; Marks, H.; Bock, C.; Gnirke, A.; Meissner, A.; et al. Sequential ChIP-bisulfite sequencing enables direct genome-scale investigation of chromatin and DNA methylation cross-talk. *Genome Res.* **2012**, *22*, 1128–1138. [CrossRef] [PubMed]
90. Galupa, R.; Heard, E. X-chromosome inactivation: New insights into cis and trans regulation. *Curr. Opin. Genet. Dev.* **2015**, *31*, 57–66. [CrossRef] [PubMed]
91. Wang, X.; Paucek, R.D.; Gooding, A.R.; Brown, Z.Z.; Ge, E.J.; Muir, T.W.; Cech, T.R. Molecular analysis of PRC2 recruitment to DNA in chromatin and its inhibition by RNA. *Nat. Struct. Mol. Biol.* **2017**, *24*, 1028–1038. [CrossRef] [PubMed]
92. Holoch, D.; Margueron, R. Mechanisms Regulating PRC2 Recruitment and Enzymatic Activity. *Trends Biochem. Sci.* **2017**, *42*, 531–542. [CrossRef] [PubMed]

93. Li, Y.; Zheng, H.; Wang, Q.; Zhou, C.; Wei, L.; Liu, X.; Zhang, W.; Zhang, Y.; Du, Z.; Wang, X.; et al. Genome-wide analyses reveal a role of Polycomb in promoting hypomethylation of DNA methylation valleys. *Genome Biol.* **2018**, *19*, 18. [CrossRef] [PubMed]
94. Tamaru, H.; Selker, E.U. A histone H3 methyltransferase controls DNA methylation in *Neurospora crassa*. *Nature* **2001**, *414*, 277–283. [CrossRef] [PubMed]
95. Arand, J.; Spieler, D.; Karius, T.; Branco, M.R.; Meilinger, D.; Meissner, A.; Jenuwein, T.; Xu, G.; Leonhardt, H.; Wolf, V.; et al. In vivo control of CpG and non-CpG DNA methylation by DNA methyltransferases. *PLoS Genet.* **2012**, *8*, e1002750. [CrossRef] [PubMed]
96. Lehnertz, B.; Ueda, Y.; Derijck, A.A.; Braunschweig, U.; Perez-Burgos, L.; Kubicek, S.; Chen, T.; Li, E.; Jenuwein, T.; Peters, A.H. Suv39h-mediated histone H3 lysine 9 methylation directs DNA methylation to major satellite repeats at pericentric heterochromatin. *Curr. Biol.* **2003**, *13*, 1192–1200. [CrossRef]
97. Rothbart, S.B.; Krajewski, K.; Nady, N.; Tempel, W.; Xue, S.; Badeaux, A.I.; Barsyte-Lovejoy, D.; Martinez, J.Y.; Bedford, M.T.; Fuchs, S.M.; et al. Association of UHRF1 with methylated H3K9 directs the maintenance of DNA methylation. *Nat. Struct. Mol. Biol.* **2012**, *19*, 1155–1160. [CrossRef] [PubMed]
98. Sharif, J.; Muto, M.; Takebayashi, S.; Suetake, I.; Iwamatsu, A.; Endo, T.A.; Shinga, J.; Mizutani-Koseki, Y.; Toyoda, T.; Okamura, K.; et al. The SRA protein Np95 mediates epigenetic inheritance by recruiting Dnmt1 to methylated DNA. *Nature* **2007**, *450*, 908–912. [CrossRef] [PubMed]
99. Bostick, M.; Kim, J.K.; Estève, P.O.; Clark, A.; Pradhan, S.; Jacobsen, S.E. UHRF1 plays a role in maintaining DNA methylation in mammalian cells. *Science* **2007**, *317*, 1760–1764. [CrossRef] [PubMed]
100. Berkyurek, A.C.; Suetake, I.; Arita, K.; Takeshita, K.; Nakagawa, A.; Shirakawa, M.; Tajima, S. The DNA methyltransferase Dnmt1 directly interacts with the SET and RING finger-associated (SRA) domain of the multifunctional protein Uhrf1 to facilitate accession of the catalytic center to hemi-methylated DNA. *J. Biol. Chem.* **2014**, *289*, 379–386. [CrossRef] [PubMed]
101. Bashtrykov, P.; Rajavelu, A.; Hackner, B.; Ragozin, S.; Carell, T.; Jeltsch, A. Targeted mutagenesis results in an activation of DNA methyltransferase 1 and confirms an autoinhibitory role of its RFTS domain. *Chembiochem* **2014**, *15*, 743–748. [CrossRef] [PubMed]
102. Xie, S.; Jakoncic, J.; Qian, C. UHRF1 double Tudor domain and the adjacent PHD finger act together to recognize K9me3-containing histone H3 tail. *J. Mol. Biol.* **2012**, *415*, 318–328. [CrossRef] [PubMed]
103. Rothbart, S.B.; Dickson, B.M.; Ong, M.S.; Krajewski, K.; Houliston, S.; Kireev, D.B.; Arrowsmith, C.H.; Strahl, B.D. Multivalent histone engagement by the linked tandem Tudor and PHD domains of UHRF1 is required for the epigenetic inheritance of DNA methylation. *Genes Dev.* **2013**, *27*, 1288–1298. [CrossRef] [PubMed]
104. Nady, N.; Lemak, A.; Walker, J.R.; Avvakumov, G.V.; Kareta, M.S.; Achour, M.; Xue, S.; Duan, S.; Allali-Hassani, A.; Zuo, X.; et al. Recognition of multivalent histone states associated with heterochromatin by UHRF1 protein. *J. Biol. Chem.* **2011**, *286*, 24300–24311. [CrossRef] [PubMed]
105. Zhao, Q.; Zhang, J.; Chen, R.; Wang, L.; Li, B.; Cheng, H.; Duan, X.; Zhu, H.; Wei, W.; Li, J.; et al. Dissecting the precise role of H3K9 methylation in crosstalk with DNA maintenance methylation in mammals. *Nat. Commun.* **2016**, *7*, 12464. [CrossRef] [PubMed]
106. Veland, N.; Hardikar, S.; Zhong, Y.; Gayatri, S.; Dan, J.; Strahl, B.D.; Rothbart, S.B.; Bedford, M.T.; Chen, T. The arginine methyltransferase PRMT6 regulates DNA methylation and contributes to global DNA hypomethylation in cancer. *Cell Rep.* **2017**, *21*, 3390–3397. [CrossRef] [PubMed]
107. Jorgensen, S.; Schotta, G.; Sorensen, C.S. Histone H4 lysine 20 methylation: Key player in epigenetic regulation of genomic integrity. *Nucleic Acids Res.* **2013**, *41*, 2797–2806. [CrossRef] [PubMed]
108. Zhu, H.; Geiman, T.M.; Xi, S.; Jiang, Q.; Schmidtmann, A.; Chen, T.; Li, E.; Muegge, K. Lsh is involved in de novo methylation of DNA. *EMBO J.* **2006**, *25*, 335–345. [CrossRef] [PubMed]
109. Yu, W.; McIntosh, C.; Lister, R.; Zhu, I.; Han, Y.; Ren, J.; Landsman, D.; Lee, E.; Briones, V.; Terashima, M.; et al. Genome-wide DNA methylation patterns in LSH mutant reveals de-repression of repeat elements and redundant epigenetic silencing pathways. *Genome Res.* **2014**, *24*, 1613–1623. [CrossRef] [PubMed]
110. Lungu, C.; Muegge, K.; Jeltsch, A.; Jurkowska, R.Z. An ATPase-deficient variant of the SNF2 family member HELLS shows altered dynamics at pericentromeric heterochromatin. *J. Mol. Biol.* **2015**, *427*, 1903–1915. [CrossRef] [PubMed]
111. Ren, J.; Briones, V.; Barbour, S.; Yu, W.; Han, Y.; Terashima, M.; Muegge, K. The ATP binding site of the chromatin remodeling homolog Lsh is required for nucleosome density and de novo DNA methylation at repeat sequences. *Nucleic Acids Res.* **2015**, *43*, 1444–1455. [CrossRef] [PubMed]

112. Song, J.; Teplova, M.; Ishibe-Murakami, S.; Patel, D.J. Structure-based mechanistic insights into DNMT1-mediated maintenance DNA methylation. *Science* **2012**, *335*, 709–712. [CrossRef] [PubMed]
113. Zhang, Z.M.; Lu, R.; Wang, P.; Yu, Y.; Chen, D.; Gao, L.; Liu, S.; Ji, D.; Rothbart, S.B.; Wang, Y.; et al. Structural basis for DNMT3A-mediated de novo DNA methylation. *Nature* **2018**, *554*, 387–391. [CrossRef] [PubMed]
114. Okuwaki, M.; Verreault, A. Maintenance DNA methylation of nucleosome core particles. *J. Biol. Chem.* **2004**, *279*, 2904–2912. [CrossRef] [PubMed]
115. Gowher, H.; Stockdale, C.J.; Goyal, R.; Ferreira, H.; Owen-Hughes, T.; Jeltsch, A. De novo methylation of nucleosomal DNA by the mammalian Dnmt1 and Dnmt3A DNA methyltransferases. *Biochemistry* **2005**, *44*, 9899–9904. [CrossRef] [PubMed]
116. Takeshima, H.; Suetake, I.; Shimahara, H.; Ura, K.; Tate, S.; Tajima, S. Distinct DNA methylation activity of Dnmt3a and Dnmt3b towards naked and nucleosomal DNA. *J. Biochem.* **2006**, *139*, 503–515. [CrossRef] [PubMed]
117. Felle, M.; Hoffmeister, H.; Rothammer, J.; Fuchs, A.; Exler, J.H.; Längst, G. Nucleosomes protect DNA from DNA methylation in vivo and in vitro. *Nucleic Acids Res.* **2011**, *39*, 6956–6969. [CrossRef] [PubMed]
118. Schrader, A.; et al. Characterization of Dnmt1 binding and DNA methylation on nucleosomes and nucleosomal arrays. *PLoS ONE* **2015**, *10*, e0140076. [CrossRef] [PubMed]
119. Xu, G.L.; Bestor, T.H.; Bourc'his, D.; Hsieh, C.L.; Tommerup, N.; Bugge, M.; Hulten, M.; Qu, X.; Russo, J.J.; Viegas-Péquignot, E. Chromosome instability and immunodeficiency syndrome caused by mutations in a DNA methyltransferase gene. *Nature* **1999**, *402*, 187–191. [CrossRef] [PubMed]
120. Okano, M.; Bell, D.W.; Haber, D.A.; Li, E. DNA methyltransferases Dnmt3a and Dnmt3b are essential for de novo methylation and mammalian development. *Cell* **1999**, *99*, 247–257. [CrossRef]
121. De Greef, J.C.; Wang, J.; Balog, J.; den Dunnen, J.T.; Frants, R.R.; Straasheijm, K.R.; Aytekin, C.; van der Burg, M.; Duprez, L.; Ferster, A.; et al. Mutations in ZBTB24 are associated with immunodeficiency, centromeric instability, and facial anomalies syndrome type 2. *Am. J. Hum. Genet.* **2011**, *88*, 796–804. [CrossRef] [PubMed]
122. Thijssen, P.E.; Ito, Y.; Grillo, G.; Wang, J.; Velasco, G.; Nitta, H.; Unoki, M.; Yoshihara, M.; Suyama, M.; Sun, Y.; et al. Mutations in CDCA7 and HELLS cause immunodeficiency-centromeric instability-facial anomalies syndrome. *Nat. Commun.* **2015**, *6*, 7870. [CrossRef] [PubMed]
123. Wu, T.P.; Wang, T.; Seetin, M.G.; Lai, Y.; Zhu, S.; Lin, K.; Liu, Y.; Byrum, S.D.; Mackintosh, S.G.; Zhong, M.; et al. DNA methylation on $N(6)$-adenine in mammalian embryonic stem cells. *Nature* **2016**, *532*, 329–333. [CrossRef] [PubMed]
124. Jenness, C.; Giunta, S.; Müller, M.M.; Kimura, H.; Muir, T.W.; Funabiki, H. HELLS and CDCA7 comprise a bipartite nucleosome remodeling complex defective in ICF syndrome. *Proc. Natl. Acad. Sci. USA* **2018**, *115*, E876–E885. [CrossRef] [PubMed]
125. Bashtrykov, P.; Ragozin, S.; Jeltsch, A. Mechanistic details of the DNA recognition by the Dnmt1 DNA methyltransferase. *FEBS Lett.* **2012**, *586*, 1821–1823. [CrossRef] [PubMed]
126. He, Y.; Ecker, J.R. Non-CG methylation in the human genome. *Annu. Rev. Genomics Hum. Genet.* **2015**, *16*, 55–77. [CrossRef] [PubMed]
127. Lee, J.H.; Park, S.J.; Nakai, K. Differential landscape of non-CpG methylation in embryonic stem cells and neurons caused by DNMT3s. *Sci. Rep.* **2017**, *7*, 11295. [CrossRef] [PubMed]
128. Kinde, B.; Gabel, H.W.; Gilbert, C.S.; Griffith, E.C.; Greenberg, M.E. Reading the unique DNA methylation landscape of the brain: Non-CpG methylation, hydroxymethylation, and MeCP2. *Proc. Natl. Acad. Sci. USA* **2015**, *112*, 6800–6806. [CrossRef] [PubMed]
129. Kinde, B.; Wu, D.Y.; Greenberg, M.E.; Gabel, H.W. DNA methylation in the gene body influences MeCP2-mediated gene repression. *Proc. Natl. Acad. Sci. USA* **2016**, *113*, 15114–15119. [CrossRef] [PubMed]
130. Cheng, J.; Yang, H.; Fang, J.; Ma, L.; Gong, R.; Wang, P.; Li, Z.; Xu, Y. Molecular mechanism for USP7-mediated DNMT1 stabilization by acetylation. *Nat. Commun.* **2015**, *6*, 7023. [CrossRef] [PubMed]
131. Keown, C.L.; Berletch, J.B.; Castanon, R.; Nery, J.R.; Disteche, C.M.; Ecker, J.R.; Mukamel, E.A. Allele-specific non-CG DNA methylation marks domains of active chromatin in female mouse brain. *Proc. Natl. Acad. Sci. USA* **2017**, *114*, E2882–E2890. [CrossRef] [PubMed]
132. O'Brown, Z.K.; Greer, E.L. N6-Methyladenine: A conserved and dynamic DNA mark. *Adv. Exp. Med. Biol.* **2016**, *945*, 213–246. [PubMed]
133. Greer, E.L.; Blanco, M.A.; Gu, L.; Sendinc, E.; Liu, J.; Aristizábal-Corrales, D.; Hsu, C.H.; Aravind, L.; He, C.; Shi, Y. DNA methylation on N6-Adenine in *C. elegans*. *Cell* **2015**, *161*, 868–878. [CrossRef] [PubMed]

134. Zhang, G.; Huang, H.; Liu, D.; Cheng, Y.; Liu, X.; Zhang, W.; Yin, R.; Zhang, D.; Zhang, P.; Liu, J.; et al. N6-methyladenine DNA modification in *Drosophila*. *Cell* **2015**, *161*, 893–906. [CrossRef] [PubMed]
135. Koziol, M.J.; Bradshaw, C.R.; Allen, G.E.; Costa, A.S.H.; Frezza, C.; Gurdon, J.B. Identification of methylated deoxyadenosines in vertebrates reveals diversity in DNA modifications. *Nat. Struct. Mol. Biol.* **2016**, *23*, 24–30. [CrossRef] [PubMed]
136. Xiao, C.L.; Zhu, S.; He, M.; Chen, D.; Zhang, Q.; Chen, Y.; Yu, G.; Liu, J.; Xie, S.Q.; Luo, F.; et al. N^6-methyladenine DNA modification in the human genome. *Mol. Cell* **2018**, *71*, 306–318. [CrossRef] [PubMed]
137. Schiffers, S.; Ebert, C.; Rahimoff, R.; Kosmatchev, O.; Steinbacher, J.; Bohne, A.V.; Spada, F.; Michalakis, S.; Nickelsen, J.; Müller, M.; et al. Quantitative LC-MS provides no evidence for m^6 dA or m^4 dC in the genome of mouse embryonic stem cells and tissues. *Angew. Chem. Int. Ed. Engl.* **2017**, *56*, 11268–11271. [CrossRef] [PubMed]
138. Van der Wijst, M.G.; Rots, M.G. Mitochondrial epigenetics: An overlooked layer of regulation? *Trends Genet.* **2015**, *31*, 353–356. [CrossRef] [PubMed]
139. Saini, S.K.; Mangalhara, K.C.; Prakasam, G.; Bamezai, R.N.K. DNA Methyltransferase1 (DNMT1) Isoform3 methylates mitochondrial genome and modulates its biology. *Sci. Rep.* **2017**, *7*, 1525. [CrossRef] [PubMed]
140. Van der Wijst, M.G.; van Tilburg, A.Y.; Ruiters, M.H.; Rots, M.G. Experimental mitochondria-targeted DNA methylation identifies GpC methylation, not CpG methylation, as potential regulator of mitochondrial gene expression. *Sci. Rep.* **2017**, *7*, 177. [CrossRef] [PubMed]
141. Liu, B.; Du, Q.; Chen, L.; Fu, G.; Li, S.; Fu, L.; Zhang, X.; Ma, C.; Bin, C. CpG methylation patterns of human mitochondrial DNA. *Sci. Rep.* **2016**, *6*, 23421. [CrossRef] [PubMed]
142. Mechta, M.; Ingerslev, L.R.; Fabre, O.; Picard, M.; Barrès, R. Evidence suggesting absence of mitochondrial DNA methylation. *Front. Genet.* **2017**, *8*, 166. [CrossRef] [PubMed]
143. Kungulovski, G.; Jeltsch, A. Epigenome editing: State of the art, concepts, and perspectives. *Trends Genet.* **2016**, *32*, 101–113. [CrossRef] [PubMed]
144. Rots, M.G.; Jeltsch, A. Editing the epigenome: Overview, open questions, and directions of future development. *Methods Mol. Biol.* **2018**, *1767*, 3–18. [PubMed]

© 2018 by the authors. Licensee MDPI, Basel, Switzerland. This article is an open access article distributed under the terms and conditions of the Creative Commons Attribution (CC BY) license (http://creativecommons.org/licenses/by/4.0/).

Review

Mechanisms of DNA Methyltransferase Recruitment in Mammals

Marthe Laisné [1], Nikhil Gupta [1], Olivier Kirsh [1], Sriharsa Pradhan [2] and Pierre-Antoine Defossez [1,*]

[1] Epigenetics and Cell Fate, UMR7216 CNRS, University Paris Diderot, Sorbonne Paris Cité, 75013 Paris, France; marthe.laisne@gmail.com (M.L.); nikhilsspp@gmail.com (N.G.); olivier.kirsh@univ-paris-diderot.fr (O.K.)
[2] New England Biolabs, 240 County Rd, Ipswich, MA 01938, USA; pradhan@neb.com
* Correspondence: pierre-antoine.defossez@univ-paris-diderot.fr

Received: 16 November 2018; Accepted: 5 December 2018; Published: 10 December 2018

Abstract: DNA methylation is an essential epigenetic mark in mammals. The proper distribution of this mark depends on accurate deposition and maintenance mechanisms, and underpins its functional role. This, in turn, depends on the precise recruitment and activation of de novo and maintenance DNA methyltransferases (DNMTs). In this review, we discuss mechanisms of recruitment of DNMTs by transcription factors and chromatin modifiers—and by RNA—and place these mechanisms in the context of biologically meaningful epigenetic events. We present hypotheses and speculations for future research, and underline the fundamental and practical benefits of better understanding the mechanisms that govern the recruitment of DNMTs.

Keywords: epigenetics; DNA methylation; DNA methyltransferases

1. DNA Methylation: An Essential and Dynamic Epigenetic Mark

The activity of the genome, especially gene expression, is regulated by epigenetic marks. This regulation has to combine two seemingly incompatible objectives: first, the epigenetic marks should be stable enough to contribute to a cell identity that is maintained through the lifetime of the cell, and that is passed on to the daughter cell [1]. Second, the marks have to be flexible enough to allow plasticity [2]. This plasticity can be very local, for instance at the scale of one promoter when a gene is induced by a stimulus. It can also be global, during the large reprogramming events that occur in the zygote after fertilization, or in primordial germ cells as they erase their parental identity to be able to produce gametes [3].

Understanding the stability and the dynamics of epigenetic marks is therefore important to discover the fundamentals of genome activity; and for detection and potential correction of the epigenetic drift that accompanies aging, as well as abnormal epigenetic reprogramming that is an underlying cause of many diseases such as cancers [4]. Finally, from a practical standpoint, understanding the stability and dynamics of chromatin marks is useful to reprogram the epigenome [5]. Again, this can be done on a single-gene scale, or on whole-genome scale to reprogram cells, which constitutes the starting point of regenerative medicine [6].

DNA methylation is a chromatin mark that is essential in mammals. It can rightly be called an "epigenetic" mark, as it has been proven to pass from mother to daughter cells, and sometimes even from one organismal generation to the next [7,8]. The main type of DNA methylation observed in mammals is the methylation of position five of cytosine within a CpG dinucleotide. Non-CpG methylation does occur, for instance in the brain [9,10], but its dynamics and roles are less understood and will not be discussed further here.

In differentiated mammalian cells, about 80% of the CpGs in the genome are methylated; however, there are marked local differences (Figure 1). Intragenic regions and repeated elements are generally methylated. CpG islands, which are associated with the promoters of about two thirds of mammalian genes, are generally unmethylated; conversely methylated promoters are often silenced. Enhancers can also be dynamically methylated, which modifies their ability to recruit transcription factors and activate transcription [11–13]. Finally, the body of actively transcribed genes is methylated, and there is a positive correlation between expression and gene body methylation [14].

Figure 1. Genome-wide distribution of DNA methylation in mammalian species. DNA methylation occurs symmetrically on CpG sites: to simplify, only one DNA strand is shown on this figure. The mammalian genome has a low frequency of CpGs, but the majority of these are methylated (black lollipops) in intergenic regions, repeated elements and transposable elements, and in gene bodies. Conversely, CpG islands are rich in CpGs and are usually protected from DNA methylation (white lollipops). Unmethylated CpG islands frequently correspond to active promoters (green arrow).

Global DNA methylation events

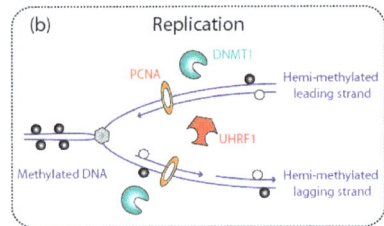

Local DNA methylation events

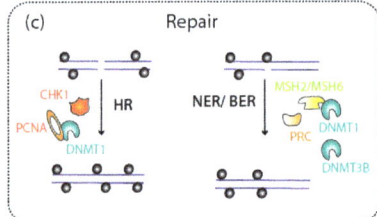
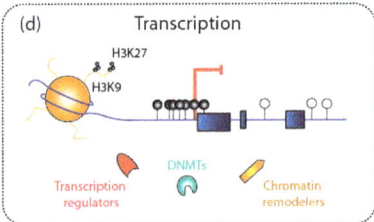

Figure 2. Biological processes involving DNA methylation, DNA methyltransferases (DNMTs) and keys partners. DNA methylation is involved in regulating many cellular processes; for all of them, DNMTs cooperate with several essential actors. (**a**,**b**) These DNA methylation events can affect the whole genome, as during the two methylation waves during mammalian life cycle, but also after DNA replicates at each cell cycle. (**c**,**d**) DNA methylation events can also be localized, as on the newly synthesized DNA after DNA repair. Selected examples, depicted in this cartoon, will be discussed further in this article. Abbreviations used: PGC: Progenitor germ cells; HR: Homologous Recombination; NER: Nucleotide Excision Repair; BER: Base Excision Repair.

DNA methylation is profoundly remodeled at several steps in the life of a mammalian organism (Figure 2). These steps include early development (Figure 2a), where the zygote after fertilization undergoes massive DNA demethylation, followed by widespread remethylation [15]. Another global demethylation wave occurs in primordial germ cells, as they erase parental imprints, before re-establishing the DNA methylation pattern found in gametes [3] (Figure 2a). The DNA methylation patterns are also globally challenged at each round of DNA replication (Figure 2b), as the newly synthesized DNA contains only unmethylated cytosines, which then have to be methylated if the parental pattern is to be maintained. Besides these genome-wide transitions, more local events are also observed: when DNA is damaged and repaired, the newly synthesized DNA is initially free of DNA methylation (Figure 2c). Also, local methylation on promoters occurs in the course of transcriptional regulation during development, or in response to a specific stimulus (Figure 2d).

The objective of this review is to summarize and discuss some of the mechanisms that are responsible for the stability and dynamics of DNA methylation, and therefore its functions. For the mark to occur, specific enzymes—the DNA methyltransferases (DNMTs) [16]—have to be recruited to target loci and become catalytically active. In this review, we will describe and discuss recent work on protein- and RNA-mediated recruitment of DNMTs, with a special emphasis on the mammalian enzymes, in the context of diverse functions DNA methylation plays in cellular processes and development.

2. Organization of the DNMTs and Its Functional Consequences

In this section, we will present an overview of the mammalian DNMTs: their domain organization, functions, and potential interacting regions with protein or RNA. The biological role of these interactions will be discussed in subsequent sections.

2.1. Several Non-Redundant Mammalian DNMTs Catalyze CpG Methylation

Cytosine methylation results from the covalent transfer of a methyl group from S-adenosyl methionine (SAM) to the carbon C-5 of cytosines to produce 5-methylcytosine (Figure 3). This activity is present in bacterial proteins, such as M. HhaI, as part of their restriction/modification systems [17], and iterative searches for mammalian proteins containing a domain similar to the bacterial enzymes led to the identification of the different DNMTs: DNMT1, DNMT2, DNMT3A, DNMT3B, DNMT3C, and DNMT3L (Figures 4 and 5). DNMT3C is present only in rodents, whereas all mammals express the other proteins.

In spite of its similarity to DNA-modifying enzymes, DNMT2 proved to be a tRNA methyltransferase [18,19] and will not be discussed further. DMNT3C is specific to muroids, and was discovered very recently [20]. Our review will for the most part focus on the better-known, catalytically active, enzymes DNMT1, DNMT3A and DNMT3B. We will also discuss DNMT3L: this protein, even though it has no intrinsic catalytic activity, is necessary to stimulate the action of DNMT3A and DNMT3B [21].

Genetic studies in the mouse [22] and other organisms showed that the DNMTs are non-redundant. One reason for their uniqueness is a specific expression pattern [22], but other factors are also involved, as the enzymes have clearly different activities in vitro [23]. Remarkably, papers by Riggs [24] as well as Holliday and Pugh [25] proposed as early as 1975 that "maintenance" DNA methylation might be distinguished from "de novo" DNA methylation, and carried out by different proteins. These predictions were validated by experimental work in the subsequent decades: broadly speaking, the maintenance of DNA methylation on hemimethylated CpG sites (generated by DNA replication), is mostly due to DNMT1, which is expressed in all cycling cells; in contrast de novo DNA methylation on both strands of previously unmethylated CpG is mostly carried out by DNMT3A and DNMT3B (Figure 2). There are exceptions to this general division of labor [26,27], but this working model is useful.

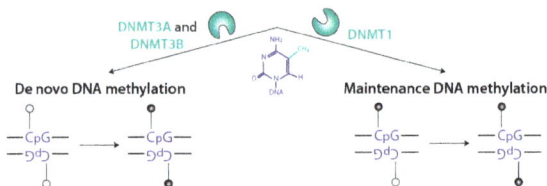

Figure 3. Specificities of the DNMTs DNA methylation can take place in two different contexts, and is processed by two distinct mechanisms: de novo DNA methylation involves fully unmethylated DNA and occurs on both strands, whereas maintenance of DNA methylation involves hemimethylated DNA.

2.2. DNMTs Have Divergent Non-Catalytic Domains

As said above, all vertebrate DNMTs share a conserved region necessary for catalysis; it permits the binding to SAM, and the recognition of DNA. This domain resembles that found in bacterial DNMTs such as M. HhaI, emphasizing their shared evolutionary origin (Figures 4 and 5).

Besides this conserved catalytic domain, which is always found at the C-terminus, the different DNMTs contain divergent N-termini that differ in size, as well as in the number and nature of domains they contain. These N-terminal regions contribute to the non-redundant functions of DNMTs.

DNMT1 is the largest of the enzymes (1616 amino acids in humans). Its catalytic domain is separated from the large N-terminal regulatory region by a series of Lys-Gly dipeptide repeats. The N-terminal part harbors different domains: (i) a charge-rich (C-R) domain, which includes a proliferating cell nuclear antigen (PCNA) binding domain (PBD); (ii) an "intrinsically disordered domain", found only in eutherian mammals [28,29], which contains a nuclear localization sequence (NLS); (iii) a replication foci target sequence (RFTS) domain; (iv) a zinc finger DNA binding domain (CXXC); (v) two bromo-adjacent homology domains (BAH1/2) [30].

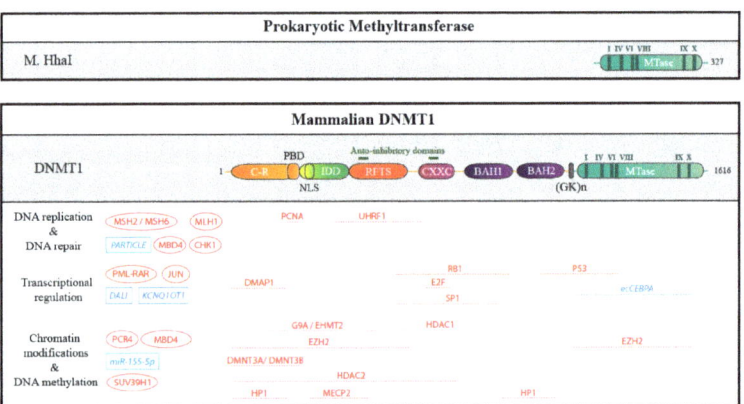

Figure 4. Schematic structure of the prokaryotic methyltransferase M. HhaI, compare to DNMT1 and partners. The human DNMT1 contains 1616 amino acids residues. The catalytic methyltransferase domain (MTase, in blue) is very similar to that of the prokaryotic methyltransferase M. HhaI and harbors highly conserved motifs (I-X, in dark blue). In addition, DNMT1 harbors a charge-rich (C-R) domain containing the proliferating cell nuclear antigen (PCNA) binding domain (PBD), an intrinsically disordered domain (IDD) with a nuclear localization sequence (NLS), a replication foci target sequence (RFTS), a zinc finger domain (CXXC), and two bromo-adjacent homology domains (BAH 1/2). The catalytic and the regulatory domains are connected by a series of Gly-Lys repeats. Auto-inhibitory domains are highlighted in green. In addition, some interacting proteins and RNAs are represented: if they are known, mapped interaction domains are indicated. Partners with unknown binding sites are shown on the left. Proteins are depicted in red, RNAs in blue.

Enzymes in the DNMT3 family have closely related architectures: DNMT3A and DNMT3B are composed of (i) the N-terminal domain, which is essential for DNA-binding; (ii) a PWWP domain which recognizes H3K36me3; (iii) an ADD-PHD domain (ATRX-DNMT3B-DNMT3L/plant homeodomain), which binds unmethylated H3K4; and (iv) the catalytic domain (Figure 5). Several isoforms of DNMT3A and DNMT3B, due to alternative promoters or splicing events, have been identified both in human and mouse, and could be involved in different functions. For example, the two major isoforms of DNMT3A shown in Figure 5 have different genomic localizations despite their high similarity [31]. The rodent-specific protein DMNT3C is similar to DNMT3B but lacks the PWWP domain [20]. DNMT3L resembles DMNT3C.

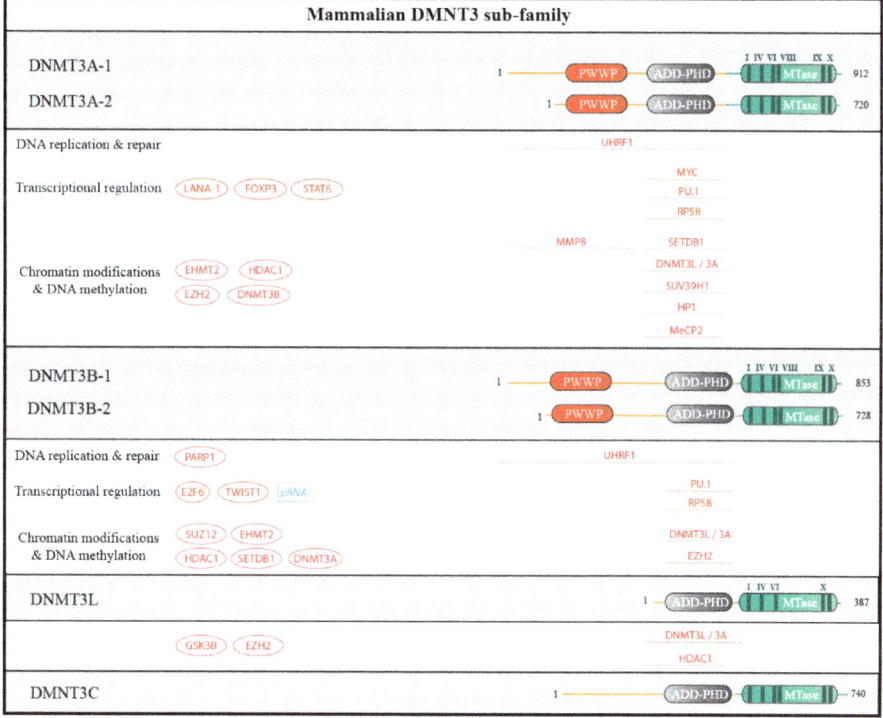

Figure 5. Schematic structure of DNMT3 sub-family and partners. Human DNMT3A, DNMT3B, DNMT3L and the rodent DNMT3C contain 912, 953, 387 and 740 amino acids residues, respectively and for the longest isoforms. For DNMT3A and DNMT3B, transcriptional isoforms due to alternative promoters are shown. The catalytic methyltransferase domain (MTase, in blue) harbors highly conserved motifs (I-X, in dark blue). DNMT3A and DNMT3B comprise a PWWP domain and an ATRX-DNMTB-DNMT3L and Plant-Homeodomain (ADD-PHD), that is also found in DNMT3L and DNMT3C. Interacting proteins and RNA are depicted as in Figure 4.

2.3. DNMTs Form Complexes

Many of the motifs described above mediate protein-protein interactions, allowing the various DNMTs to participate in multiprotein complexes.

DNMT3A can form hetero-oligomeric complexes: as a heteroduplex with DNMT3L, which increases the processivity of the enzyme, or as a linear heterotetramer with two DNMT3L subunits (at the edges of the tetramer) and two DNMT3A subunits [32]. Moreover, DNMT3A alone or DNMT3A/DNMT3L complexes can also cooperatively bind DNA, and form large multimeric DNA/protein fibers [33].

Besides homo- or hetero-oligomerization, biochemical approaches such as immunoprecipitation followed by mass spectrometry have revealed that the DNMTs, as other chromatin-modifying enzymes, are part of larger protein complexes [34–36]. Estimating the stoichiometry of these complexes can help reveal which members of the complex are constitutively associated to the DNMTs, and which are minor or transient interactors; this quantitative approach has been historically challenging, but emerging technologies should help improve the situation [37].

2.4. DNMTs Bind Nucleic Acids

The DNMTs contain several domains that can bind nucleic acids, i.e., DNA or RNA. The CXXC domain of DNMT1 is a type of zinc-finger with preferred binding to unmethylated (rather than methylated) CpGs, with an important role in the auto-inhibition mechanism [30]. The catalytic domain also contains a number of basic residues which interact in a sequence-independent manner with the negatively charged DNA backbone [38]. Interestingly, the DNMTs have an affinity for G-quadruplexes [39], and this is biologically relevant [40]. Finally, the PWWP of the DNMT3s has been shown to bind DNA [41].

2.5. DNMTs Are Autoinhibited

Finally, an important functional characteristic of the DNMTs is that they are intramolecularly inhibited, which presumably decreases their off-target activity. Structural and biochemical experiments have that DNMT1 is inhibited by intramolecular interactions between the catalytic site and the RFTD domain or the CXXC domain [5]. DNMT3A is also autoinhibited, albeit by a different domain, the ADD [42,43].

2.6. Functional Consequences for Recruitment Mechanisms

Four important conclusions can be drawn from this overview of the DNMTs. First, the enzymes have a large number of domains, structured or unstructured, with which to establish protein-protein or protein-nucleic acid interactions. Second, some of these domains engage in intramolecular interactions with the catalytic domain and inhibit its activity. Therefore, the recruitment of DNMTs by a protein or RNA interactor may have two separate effects: increasing the local enzyme concentration, but also activating the enzyme at its site of recruitment [42]. Third, the recruitment of a DNMT will not necessarily lead to local methylation: the interaction could in fact break up a catalytically productive complex, or stabilize the auto-inhibited form of the enzyme [44]. Fourth, while DNA methylation is of course the best-known activity of DNMTs, they might possess other important functions that are unrelated to DNA methylation. These functions could be intrinsic, or borne by DNMT interactors: it is clear, for instance, that DNMTs associate with other chromatin-modifying factors, such as histone deacetylases (HDACs) [45]. Therefore, recruiting a DNMT may alter DNA methylation locally, but it could also have other consequences on chromatin.

3. DNMT Recruitment in the Regulation of Chromatin Structure and Gene Expression

DNA methylation is deeply linked to cell identity, as it is a determinant of the cellular transcriptional program [46]. Besides regulating cellular gene expression, DNA methylation is also a key contributor to the transcriptional repression of transposons [21]. These functions of DNA methylation depend on the recruitment of DNMTs with transcription factors and other DNA binding proteins, with histone marks and chromatin modifiers, and with non-coding RNAs. We review these interactions in the subsequent sections, with an emphasis on the most recent data.

3.1. Interaction with Promoter-Bound Transcription Factors

It is well described that DNA methylation status can influence the recruitment of transcriptional regulators [47]. Conversely, transcription factors bound to DNA can also directly recruit the DNA methylation machinery. This was first reported for an oncogenic transcription factor, PML-RAR [48], but the paradigm was rapidly extended to unaltered cellular transcription factors, such as p53 recruiting DNMT1 to silence the *SURVIVIN* promoter [49], and MYC recruiting DNMT3a to silence *p21/CDKN1A* [50]. Since then, many other examples of cellular [46] or viral [47] transcription factors recruiting DNMTs to promoters, via direct interactions, have been discovered. This topic, however, has been discussed in a previous review [48]. Most reported examples concern the recruitment of DNMTs to promoter regions. Interestingly, a previous paper showed that the zinc finger protein ZBTB24, which is found mutated in Immunodeficiency, Centromere instability and Facial anomalies (ICF) syndrome, is likely to recruit DNMT3B to certain gene bodies [49]. This mechanism seems to apply not only to the genes transcribed by RNA Polymerase II (PolII), but also the genes that are targets of PolI [50] or PolIII [51] Also, while the recruitment of DNMTs seems to generally be accompanied by transcriptional repression, the mechanisms may be varied, at least for DNMT1. In some cases, methylation of DNA by the enzymes seems to be the cause of repression, while in others the enzyme may repress transcription independently of its catalytic activity [51]. This non-catalytic repression seems itself due to protein-protein interactions by which DNMT1 recruits chromatin-modifying enzymes [52].

Besides the relative contributions of catalytic and non-catalytic functions of DNMTs to promoter activity, several general questions await clarification. In a given cell, what fraction of the DNMT molecules is engaged in transcriptional regulation? What are the dynamics of these interactions, and are they regulated by modifications of the transcription factors, the DNMTs, or both? In all but a few examples [53], this regulation is unknown. Do the same mechanisms occur at enhancers [54]? Finally, how can the DNMTs interact with such a large number of unrelated transcriptional regulators? The situation is somewhat reminiscent of the transcription machinery, which can be recruited by many different, apparently unstructured transcriptional activation domains [55]. It would be of interest to determine whether the DNMTs also contain low-complexity regions that function in a similar manner.

3.2. Interaction with Chromatin Modifiers

DNMT1 has different histone-binding partners (histone-methyltransferases, histone deacetylases, but also nucleosome remodelers like SNF2H), mainly recruited through its N-terminal domain, and depicted in Figure 4. An illustrative example is the interaction between DNMT1 and the H3K9 methyltransferase G9a/EHMT2 [56], which helps coordinate DNA and histone methylation after DNA replication. More generally, this crosstalk between the DNA and H3K9 methylation pathways seems fairly prevalent [57], and may be of particular importance at repeated sequences such as centromeres [58].

Less is known about the interactome of DNMT3A and DNMT3B, but some of their chromatin-modifier partners have been identified: for example, DNMT3A interacts with the histone-lysine methyltransferase SETDB1 through its plant homeodomain (PHD) zinc finger and contributes to gene silencing [59]. The relationship between the Polycomb machinery and the DNA methylation machinery is probably more complex than initially thought [57]: while it has been ascertained that EZH2 can in fact recruit DNMT3A to the genome, this recruitment is not sufficient to trigger de novo DNA methylation [60]. The histone-binding protein MPP8 forms a molecular bridge between EHMT1/GLP and DNMT3A, which may help coordinate DNA methylation and H3K9 methylation [61]. The interaction involves the chromodomain of MPP8, which binds a methylated lysine in the N-terminus of DNMT3A [61]. A last example is that the ATRX domain of DNMT3A and the histone acetyltransferase HDAC1 can interact; consequently DNMT3A promotes histone deacetylation near the binding sites of its interactor RP58 [62].

3.3. Integrating Interaction Mechanisms to Dynamically Regulate a Transcriptional Program: The Example of Germline Genes

Genes specifically expressed in the germline permit the formation of gametes; examples of such genes are those necessary for meiosis. Their expression is tightly repressed in somatic cells, and it was recognized early on that many of these genes require DNA methylation for repression [63]. Genetic and molecular experiments showed that the methylation of germline genes is laid by DNMT3B [63], and that this deposition was crucially dependent on the transcription factor E2F6, which recruits DNMT3B to its target sites [64]. This transcription factor/DNMT interaction is one part of a complex web of regulation, as the histone modifying enzyme G9a/EHMT2 is also necessary for DNA methylation to occur on certain germline genes [65]. In addition, E2F6 also takes part in a parallel transcriptional repression mechanism, involving a non-canonical polycomb repressive complex 1 (PRC1) complex [66]. The germline genes therefore provide a clear example of the superposition of DNMT recruitment mechanisms and of repressive pathways [67] (Figure 6).

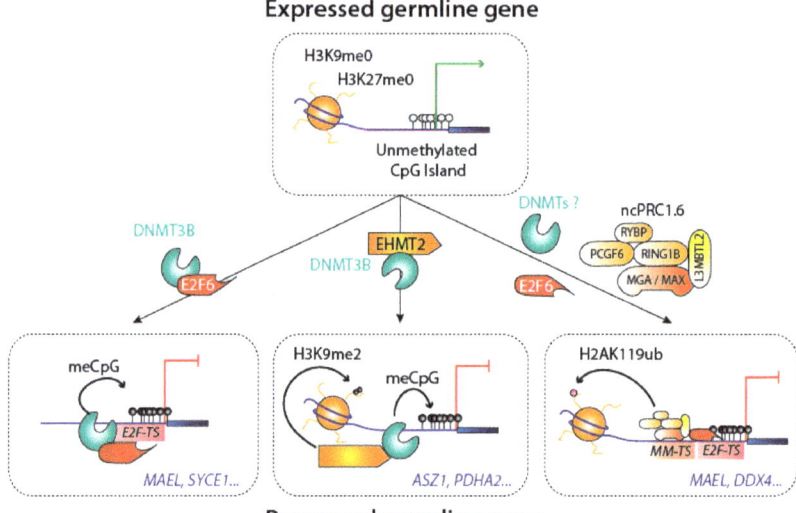

Figure 6. Silencing of the germline genes during implantation. Different complexes are involved in the silencing of germline genes. Three well documented mechanisms are shown, with some example of known targeted germline genes in blue. Abbreviations used: *E2F-TS*: E2F target sequence; *MM-TS*: MGA-MAX target sequence.

3.4. The Role of lncRNAs and miRNAs

Over the past decade, long non-coding RNAs (lncRNAs) have emerged as major regulators of the genome, and they act in part by recruiting nucleoprotein complexes [68]. Mouse and human DNMTs are among the chromatin factors that can be recruited by lncRNAs.

One illustrative instance of this principle was described at the mouse ribosomal DNA (rDNA) locus. There, promoter-associated RNAs (pRNAs), ~200 nucleotides in length, are formed and remain associated with the rDNA promoter via the formation of an RNA:DNA triplex [69]. Fascinatingly, they also associate with DNMT3B, promote its recruitment to the locus, and the inhibition of rDNA transcription [69]. To the best of our knowledge, the region of DNMT3B that is recruited by the pRNA has not been described. However, the recruitment by triplex-forming RNAs has now been shown for DNMT1 as well, as in the case of *PARTICLE*, a lncRNA induced by genotoxic insult [70,71].

Other cases of DNMT cis-recruitment by lncRNA do not seem to involve a triplex-formed RNA. For instance, the *KCNQ1* is a paradigmatic imprinted region, i.e., a region in which the alleles on the paternal and maternal chromosomes have a stable and reciprocal expression pattern. Imprinted genes often produce lncRNA, which participate in the allele-specific regulation of the locus [72]. It was observed that the antisense RNA *KCNQ1OT1*, produced by the paternal allele, interacts with DNMT1, and recruits it to the paternal chromosome, where expression is silenced [73]. The lncRNA-mediated recruitment of DNMTs may also function in trans, as shown with a regulator of neuronal development, *DALI* [74], or within the tumor-suppressive PTEN locus [75].

Although recruitment of DNMTs by RNA is thought to repress the target loci, a recent report challenges this notion [75]. In this paper, it is shown that thousands of loci produce lncRNA which associate with DNMT1. One specific lncRNA, starting upstream of *CEBPA* gene and named *ecCEBPA*, is studied in more detail, and found to adopt a stem-loop structure which binds with high affinity to the catalytic domain of DNMT1. Importantly, this lncRNA protects the *CEBPA* gene against methylation, again illustrating the principle that recruitment of DNMTs does not necessarily equate increased local DNA methylation.

It is noteworthy that microRNAs generated by the ribonuclease DROSHA have been recently shown to bind DNMT1 and decrease its activity [76]. At least one of them, *miR-155-5p*, can apparently act globally and decrease genome methylation [77]. Future work will hopefully reveal whether these mechanisms can also act in recruitment pathways.

3.5. Repression of Transposable Elements: The Role of piRNAs

During its evolution, the mammalian genome has been populated by many transposable elements (TEs). The majority of these sequences are now inactive due to accumulated mutations; however, a small number of copies still retain the potential to transpose. It is now accepted that these mobile elements have a positive role as "evolutionary drivers" [78–80]. Nevertheless, transposition is potentially harmful, and cells have evolved complex mechanisms to tightly control transposons, in part at the transcriptional level [81].

The transcriptional repression of transposons depends heavily on DNA methylation [21], and uses several protein-mediated DNMT recruitment mechanisms. For instance, repression of transposons by TRIM28 leads to DNA methylation [82], although the mechanistic details are lacking. But an original RNA-dependent recruitment pathway is also involved. Indeed, one of the most crucial mechanisms to silence the TEs during male gametogenesis is the PIWI/piRNA pathway in which the piRNAs (small RNAs of 25–32 nt) combine with the PIWI proteins from the Argonaute family to form piRISC complexes. These are directed in a RNA-directed manner to initiate the repression of TEs by recruiting histone modifiers and DNMTs.

PIWI proteins have a defined expression window, among them, MIWI2 expression coincides with the de-novo methylation wave in PGC and also has a nuclear localization, while another critical PIWI protein MILI has a broader expression window and is exclusively cytoplasmic. MIWI2 is proposed to recruit DNMTs, however a detailed mechanism for the recruitment of DNMTs is yet to be elucidated. Further, loss of function studies for MILI and MIWI2 indicate that they may have some non-overlapping roles in DNA methylation of TEs, thus highlighting unknown mechanisms for the recruitment of DNMTs at TEs [81,83,84]. Further, other studies postulate the role of piRNA mediated DNA methylation well beyond the TEs, in the regulation of mRNA transcripts in both somatic and germ cells, imprinted DMR locus and oncogenes involved in cancer [85–87].

4. Maintenance of DNA Methylation during DNA Replication

A very well described role of DNMT1 is to carry out maintenance DNA methylation following DNA replication. This is mediated by a number of well-characterized interactions, which are described in the following paragraphs.

4.1. Interaction with the DNA Replication Machinery

As noted above, it was realized early on that DNA replication would reset the genome to a hemimethylated state, and that if the marks were to be maintained stably through cell divisions, a maintenance mechanism must exist. It was therefore a momentous advance when DNMT1 was shown to directly contact PCNA, providing a direct molecular link between DNA replication and DNA methylation [88]. The exact interacting region was found by pull-down assays against different DNMT1 fragments [89]. It occurs through a conserved PCNA-interacting protein (PIP) motif, a motif which is found in many of the proteins that interact with PCNA and is usually located in intrinsically disordered regions of the proteins [90]. Interestingly, the inactivation of this motif clearly prevents the interaction of DNMT1 with PCNA, and impedes the recruitment of DNMT1 to replication foci in early and mid S phase, but it does not affect the steady-state level of DNA methylation in mouse ES cells [91]. In other words, the direct interaction with PCNA seems to facilitate the recruitment of DNMT1 to replication sites, but other mechanisms can compensate if this interaction is not permitted. We will touch on some of these mechanisms in the following sections.

4.2. Role of UHRF1 and Recognition of Modified Histones

Besides DNMT1, another protein is known to be critically required for DNA methylation maintenance: UHRF1 [92,93]. Initial models suggested that UHRF1 promotes DNA methylation maintenance by directly recruiting DNMT1 through the RFTS domain [94] and then activating the enzyme [95,96], but more recent work suggest that less direct mechanisms are also involved. In particular, UHRF1 is a ubiquitin ligase that can modify histones, and ubiquitinated histones can be bound by the RFTS domain of DNMT1 [97,98]. This important topic is covered in more detail elsewhere in this issue.

4.3. Unresolved Questions

There is no doubt that DNA methylation maintenance is, somehow, coupled to DNA replication. However, as new discoveries are made, it emerges that the simple and elegant model first reported—a direct interaction of DNMT1 with PCNA—coexists with others. Several important questions are still unanswered. The kinetics with which DNA methylation is re-established is a matter of controversy [99,100]. Are lncRNAs involved in DNA methylation maintenance? How much do the "de novo" methyltransferases contribute to maintenance activity, and how [26]? Are the mechanisms of DNA methylation maintenance identical on leading and lagging strand [101]? The answer to these questions is directly linked to the identification of mechanisms recruiting the respective enzymes to their targets.

5. Restoration of DNA Methylation after DNA Damage

After DNA damage, different types of DNA repair can take place: Nucleotide Excision Repair (NER), Base Excision Repair (BER), Non-homologous End Joining (NHEJ) and Homologous Recombination (HR). The choice depends on the type of damage (for instance, presence of adducts versus presence of a double-strand break), and on the position within the cell cycle, which determines whether a sister chromatid is present to serve as a template for repair. Some types of repair, such as HR, entail the resection and resynthesis of large sections of DNA (up to several kilobases in mammals [102]). Therefore, it is expected that the DNA methylation machinery should be recruited to sites of HR to permit the re-establishment of the mark on the newly re-synthesized DNA, and this prediction has indeed been verified experimentally for DNMT1, both by microscopy [103,104] and by biochemical assays [105,106]. It is likely that one mechanism of recruitment is the direct interaction with PCNA, as the kinetics of recruitment of PCNA and DNMT1 are similar, and a region containing the interaction motif is necessary for the recruitment to occur [103]. Nevertheless, it is possible that other recruitment mechanisms also take place; in fact direct interaction with the DNA damage response protein CHK1 has been shown, but has not been worked out in mechanistic detail [107]. Yet other mechanisms may exist: for instance, UHRF1 has been proposed to directly recognize certain types of damaged DNA [108,109], and it would be of interest to determine whether this permits the recruitment of DNMT1 in parallel to the PCNA-driven pathway.

A particular situation leading to DNA damage is oxidative stress [110]. Oxidized bases such as 8-oxo-G are repaired in large part by BER and NER, but not HR [111]. An interesting report showed that oxidative stress led to the formation of large complexes containing DNMT1, DNMT3B, and Polycomb proteins, which were then addressed to previously unmethylated CpG islands [112]. Follow-up work recently clarified the mechanism, which depends on the mismatch repair proteins MSH2/MSH6 [113,114], but not on PCNA. While the interaction of DNMT1 with MSH2/MSH6 is clearly stimulated by oxidative stress, its mechanistic underpinning remains to be precised: what are the domains of the proteins involved? Why is the interaction stimulated by oxidation? These results echo earlier findings showing an interaction of DNMT1 with the mismatch repair protein MLH1 [104].

To summarize, DNMTs are recruited to chromatin after several types of DNA damage, double-strand breaks and oxidative damage have both been proved directly, and it would be interesting to assess whether other lesions, such as single-strand breaks or pyrimidine dimers also have the same effect. One mechanism that is unambiguously involved is the interaction of DNMT1 with PCNA. Interactions with mismatch repair proteins also appear important, but their molecular basis is unclear. Needless to say, the future may reveal yet other modes of DNMT recruitment after DNA damage.

6. Conclusions and Perspectives

6.1. Conceptual Advances in the Roles of DNMTs

The development of high-throughput sequencing technologies has revolutionized our ability to map DNA methylation, to identify genomic loci bound by DNMTs, and to identify RNAs associated with DNMTs. In parallel, advances in mass spectrometry have also made it much easier to detect and quantify protein complexes, while new methods in microscopy give us insight into their location and dynamics. The combination of these methods has allowed the community to arrive at the cumulative knowledge presented in this review. Much of the evidence has cemented early insight that DNA methylation is a critical epigenetic mechanism, which contributes to cell identity by regulating transcriptional programs, by ensuring the proper chromatin composition on key chromosome elements such as centromeres, and by repressing repeated elements. Another important lesson is that DNA methylation is part of a complex mesh of chromatin regulation mechanisms, which includes non-coding RNAs, histone- and nucleosome-modifiers, and DNA demethylation activities [15]. An important issue, not discussed here, is how DNA methylation is coordinated to DNA demethylation to achieve the final patterns seen in cells [115].

6.2. Targeting the DNMTs for Epigenome Editing

Artificially recruiting DNMTs to a locus of interest has long been considered as potentially useful to "edit the epigenome", for example to turn off the expression of oncogenes in tumor cells, or to remodel the genome of stem cells [116]. The idea has been made much easier to implement with the development of Cas9-based platforms [5,117]. The results described in this review have practical applications. First, they show that non-catalytic activities of DNMTs can be critical for repression and may have to be maintained for a Cas9 fusion to work efficiently. Second, they underline that increasing the local concentration of DNMTs by recruitment does not always translate into increased DNA methylation, as self-inhibitory mechanisms have to be overcome. Third, they show that RNA can be explored as a way to recruit DNMTs. Again, the recruitment can either lead to local DNA methylation and repression [85] or, conversely, to inhibition of the enzyme and local protection from its activity [118].

6.3. Consequences for Disease and Treatment

Most of the mechanisms we have described occur during the development and life of a healthy organism. We have already mentioned, however, that some of these mechanisms can be subverted by viral or oncogenic proteins. Understanding these events molecularly may help target them pharmaceutically to fight infections and cancers. More generally, the DNA methylation patterns drift during human aging, and this may contribute to the increase in cancer risk with age [119]. It will be of great interest in the future to determine if the DNMT recruitment mechanisms that have been identified go awry in aging cells, and whether this can be prevented or reversed pharmaceutically.

Author Contributions: All authors contributed to the writing and correction of the review.

Funding: Work in the lab of P.A.D. was supported by Association pour la Recherche contre le Cancer (ARC2014), by Agence Nationale de la Recherche (ANR-15-CE12-0012-01 and ANR-11-LABX-0071 under ANR-11-IDEX-0005-01), and by Institut National du Cancer (INCa PLBio 2015-1-PLBio-01-DR A-1).

Acknowledgments: We are grateful to members of the Defossez lab for useful discussions. We apologize to the authors of relevant primary papers which could not be cited because of space constraints.

Conflicts of Interest: The authors declare no conflict of interest.

References

1. Yadav, T.; Quivy, J.-P.; Almouzni, G. Chromatin plasticity: A versatile landscape that underlies cell fate and identity. *Science* **2018**, *361*, 1332–1336. [CrossRef] [PubMed]
2. Atlasi, Y.; Stunnenberg, H.G. The interplay of epigenetic marks during stem cell differentiation and development. *Nat. Rev. Genet.* **2017**, *18*, 643–658. [CrossRef]
3. Nashun, B.; Hill, P.W.S.; Hajkova, P. Reprogramming of cell fate: Epigenetic memory and the erasure of memories past. *EMBO J.* **2015**, *34*, 1296–1308. [CrossRef] [PubMed]
4. Dawson, M.A. The cancer epigenome: Concepts, challenges, and therapeutic opportunities. *Science* **2017**, *355*, 1147–1152. [CrossRef] [PubMed]
5. Kungulovski, G.; Jeltsch, A. Epigenome editing: State of the art, concepts, and perspectives. *Trends Genet. TIG* **2016**, *32*, 101–113. [CrossRef] [PubMed]
6. Takahashi, K.; Yamanaka, S. A decade of transcription factor-mediated reprogramming to pluripotency. *Nat. Rev. Mol. Cell Biol.* **2016**, *17*, 183–193. [CrossRef] [PubMed]
7. Schübeler, D. Function and information content of DNA methylation. *Nature* **2015**, *517*, 321–326. [CrossRef]
8. Luo, C.; Hajkova, P.; Ecker, J.R. Dynamic DNA methylation: In the right place at the right time. *Science* **2018**, *361*, 1336–1340. [CrossRef]
9. Lister, R.; Mukamel, E.A.; Nery, J.R.; Urich, M.; Puddifoot, C.A.; Johnson, N.D.; Lucero, J.; Huang, Y.; Dwork, A.J.; Schultz, M.D.; et al. Global epigenomic reconfiguration during mammalian brain development. *Science* **2013**, *341*, 1237905. [CrossRef]
10. Kinde, B.; Gabel, H.W.; Gilbert, C.S.; Griffith, E.C.; Greenberg, M.E. Reading the unique DNA methylation landscape of the brain: Non-CpG methylation, hydroxymethylation, and MeCP2. *Proc. Natl. Acad. Sci. USA* **2015**, *112*, 6800–6806. [CrossRef]
11. Yao, L.; Shen, H.; Laird, P.W.; Farnham, P.J.; Berman, B.P. Inferring regulatory element landscapes and transcription factor networks from cancer methylomes. *Genome Biol.* **2015**, *16*, 105. [CrossRef] [PubMed]
12. Zhang, Y.; Zhang, D.; Li, Q.; Liang, J.; Sun, L.; Yi, X.; Chen, Z.; Yan, R.; Xie, G.; Li, W.; et al. Nucleation of DNA repair factors by FOXA1 links DNA demethylation to transcriptional pioneering. *Nat. Genet.* **2016**, *48*, 1003–1013. [CrossRef] [PubMed]
13. Fleischer, T.; Tekpli, X.; Mathelier, A.; Wang, S.; Nebdal, D.; Dhakal, H.P.; Sahlberg, K.K.; Schlichting, E.; Oslo Breast Cancer Research Consortium (OSBREAC); Børresen-Dale, A.-L.; et al. DNA methylation at enhancers identifies distinct breast cancer lineages. *Nat. Commun.* **2017**, *8*, 1379. [CrossRef] [PubMed]
14. Yang, X.; Han, H.; De Carvalho, D.D.; Lay, F.D.; Jones, P.A.; Liang, G. Gene body methylation can alter gene expression and is a therapeutic target in cancer. *Cancer Cell* **2014**, *26*, 577–590. [CrossRef] [PubMed]
15. Iurlaro, M.; von Meyenn, F.; Reik, W. DNA methylation homeostasis in human and mouse development. *Curr. Opin. Genet. Dev.* **2017**, *43*, 101–109. [CrossRef] [PubMed]

16. Gowher, H.; Jeltsch, A. Mammalian DNA methyltransferases: New discoveries and open questions. *Biochem. Soc. Trans.* **2018**, *46*, 1191–1202. [CrossRef] [PubMed]
17. Blow, M.J.; Clark, T.A.; Daum, C.G.; Deutschbauer, A.M.; Fomenkov, A.; Fries, R.; Froula, J.; Kang, D.D.; Malmstrom, R.R.; Morgan, R.D.; et al. The epigenomic landscape of prokaryotes. *PLoS Genet.* **2016**, *12*, e1005854. [CrossRef]
18. Goll, M.G.; Kirpekar, F.; Maggert, K.A.; Yoder, J.A.; Hsieh, C.-L.; Zhang, X.; Golic, K.G.; Jacobsen, S.E.; Bestor, T.H. Methylation of tRNA^{Asp} by the DNA methyltransferase homolog Dnmt2. *Science* **2006**, *311*, 395–398. [CrossRef]
19. Defossez, P.-A. Ceci n'est pas une DNMT: Recently discovered functions of DNMT2 and their relation to methyltransferase activity (Comment on DOI 10.1002/bies.201300088). *BioEssays* **2013**, *35*, 1024. [CrossRef]
20. Barau, J.; Teissandier, A.; Zamudio, N.; Roy, S.; Nalesso, V.; Hérault, Y.; Guillou, F.; Bourc'his, D. The DNA methyltransferase DNMT3C protects male germ cells from transposon activity. *Science* **2016**, *354*, 909–912. [CrossRef]
21. Edwards, J.R.; Yarychkivska, O.; Boulard, M.; Bestor, T.H. DNA methylation and DNA methyltransferases. *Epigenetics Chromatin* **2017**, *10*, 23. [CrossRef] [PubMed]
22. Dan, J.; Chen, T. Genetic studies on mammalian dna methyltransferases. *Adv. Exp. Med. Biol.* **2016**, *945*, 123–150. [CrossRef] [PubMed]
23. Ambrosi, C.; Manzo, M.; Baubec, T. Dynamics and context-dependent roles of DNA methylation. *J. Mol. Biol.* **2017**, *429*, 1459–1475. [CrossRef] [PubMed]
24. Riggs, A.D. X inactivation, differentiation, and DNA methylation. *Cytogenet. Cell Genet.* **1975**, *14*, 9–25. [CrossRef] [PubMed]
25. Holliday, R.; Pugh, J.E. DNA modification mechanisms and gene activity during development. *Science* **1975**, *187*, 226–232. [CrossRef] [PubMed]
26. Walton, E.L.; Francastel, C.; Velasco, G. Maintenance of DNA methylation: Dnmt3b joins the dance. *Epigenetics* **2011**, *6*, 1373–1377. [CrossRef]
27. Elliott, E.N.; Sheaffer, K.L.; Kaestner, K.H. The "de novo" DNA methyltransferase Dnmt3b compensates the Dnmt1-deficient intestinal epithelium. *eLife* **2016**, *5*. [CrossRef]
28. Shaffer, B.; McGraw, S.; Xiao, S.C.; Chan, D.; Trasler, J.; Chaillet, J.R. The DNMT1 intrinsically disordered domain regulates genomic methylation during development. *Genetics* **2015**, *199*, 533–541. [CrossRef] [PubMed]
29. Borowczyk, E.; Mohan, K.N.; D'Aiuto, L.; Cirio, M.C.; Chaillet, J.R. Identification of a region of the DNMT1 methyltransferase that regulates the maintenance of genomic imprints. *Proc. Natl. Acad. Sci. USA* **2009**, *106*, 20806–20811. [CrossRef]
30. Song, J.; Rechkoblit, O.; Bestor, T.H.; Patel, D.J. Structure of DNMT1-DNA complex reveals a role for autoinhibition in maintenance DNA methylation. *Science* **2011**, *331*, 1036–1040. [CrossRef]
31. Manzo, M.; Wirz, J.; Ambrosi, C.; Villaseñor, R.; Roschitzki, B.; Baubec, T. Isoform-specific localization of DNMT3A regulates DNA methylation fidelity at bivalent CpG islands. *EMBO J.* **2017**, *36*, 3421–3434. [CrossRef] [PubMed]
32. Jia, D.; Jurkowska, R.Z.; Zhang, X.; Jeltsch, A.; Cheng, X. Structure of Dnmt3a bound to Dnmt3L suggests a model for de novo DNA methylation. *Nature* **2007**, *449*, 248–251. [CrossRef] [PubMed]
33. Jurkowska, R.Z.; Anspach, N.; Urbanke, C.; Jia, D.; Reinhardt, R.; Nellen, W.; Cheng, X.; Jeltsch, A. Formation of nucleoprotein filaments by mammalian DNA methyltransferase Dnmt3a in complex with regulator Dnmt3L. *Nucleic Acids Res.* **2008**, *36*, 6656–6663. [CrossRef] [PubMed]
34. Wan, C.; Borgeson, B.; Phanse, S.; Tu, F.; Drew, K.; Clark, G.; Xiong, X.; Kagan, O.; Kwan, J.; Bezginov, A.; et al. Panorama of ancient metazoan macromolecular complexes. *Nature* **2015**, *525*, 339–344. [CrossRef]
35. Huttlin, E.L.; Bruckner, R.J.; Paulo, J.A.; Cannon, J.R.; Ting, L.; Baltier, K.; Colby, G.; Gebreab, F.; Gygi, M.P.; Parzen, H.; et al. Architecture of the human interactome defines protein communities and disease networks. *Nature* **2017**, *545*, 505–509. [CrossRef] [PubMed]
36. Ponnaluri, V.K.C.; Estève, P.-O.; Ruse, C.I.; Pradhan, S. S-adenosylhomocysteine hydrolase participates in DNA methylation inheritance. *J. Mol. Biol.* **2018**, *430*, 2051–2065. [CrossRef] [PubMed]
37. Smits, A.H.; Vermeulen, M. characterizing protein-protein interactions using mass spectrometry: Challenges and opportunities. *Trends Biotechnol.* **2016**, *34*, 825–834. [CrossRef]

38. Song, J.; Teplova, M.; Ishibe-Murakami, S.; Patel, D.J. Structure-based mechanistic insights into DNMT1-mediated maintenance DNA methylation. *Science* **2012**, *335*, 709–712. [CrossRef]
39. Cree, S.L.; Fredericks, R.; Miller, A.; Pearce, F.G.; Filichev, V.; Fee, C.; Kennedy, M.A. DNA G-quadruplexes show strong interaction with DNA methyltransferases in vitro. *FEBS Lett.* **2016**, *590*, 2870–2883. [CrossRef]
40. Mao, S.-Q.; Ghanbarian, A.T.; Spiegel, J.; Martínez Cuesta, S.; Beraldi, D.; Di Antonio, M.; Marsico, G.; Hänsel-Hertsch, R.; Tannahill, D.; Balasubramanian, S. DNA G-quadruplex structures mold the DNA methylome. *Nat. Struct. Mol. Biol.* **2018**, *25*, 951–957. [CrossRef]
41. Qiu, C.; Sawada, K.; Zhang, X.; Cheng, X. The PWWP domain of mammalian DNA methyltransferase Dnmt3b defines a new family of DNA-binding folds. *Nat. Struct. Biol.* **2002**, *9*, 217–224. [CrossRef] [PubMed]
42. Jeltsch, A.; Jurkowska, R.Z. Allosteric control of mammalian DNA methyltransferases—A new regulatory paradigm. *Nucleic Acids Res.* **2016**, *44*, 8556–8575. [CrossRef] [PubMed]
43. Guo, X.; Wang, L.; Li, J.; Ding, Z.; Xiao, J.; Yin, X.; He, S.; Shi, P.; Dong, L.; Li, G.; et al. Structural insight into autoinhibition and histone H3-induced activation of DNMT3A. *Nature* **2015**, *517*, 640–644. [CrossRef] [PubMed]
44. Rajavelu, A.; Lungu, C.; Emperle, M.; Dukatz, M.; Bröhm, A.; Broche, J.; Hanelt, I.; Parsa, E.; Schiffers, S.; Karnik, R.; et al. Chromatin-dependent allosteric regulation of DNMT3A activity by MeCP2. *Nucleic Acids Res.* **2018**, *46*, 9044–9056. [CrossRef] [PubMed]
45. Jones, P.A.; Issa, J.-P.J.; Baylin, S. Targeting the cancer epigenome for therapy. *Nat. Rev. Genet.* **2016**, *17*, 630–641. [CrossRef] [PubMed]
46. Bogdanović, O.; Lister, R. DNA methylation and the preservation of cell identity. *Curr. Opin. Genet. Dev.* **2017**, *46*, 9–14. [CrossRef]
47. Yin, Y.; Morgunova, E.; Jolma, A.; Kaasinen, E.; Sahu, B.; Khund-Sayeed, S.; Das, P.K.; Kivioja, T.; Dave, K.; Zhong, F.; et al. Impact of cytosine methylation on DNA binding specificities of human transcription factors. *Science* **2017**, *356*. [CrossRef]
48. Di Croce, L.; Raker, V.A.; Corsaro, M.; Fazi, F.; Fanelli, M.; Faretta, M.; Fuks, F.; Lo Coco, F.; Kouzarides, T.; Nervi, C.; et al. Methyltransferase recruitment and DNA hypermethylation of target promoters by an oncogenic transcription factor. *Science* **2002**, *295*, 1079–1082. [CrossRef]
49. Estève, P.-O.; Chin, H.G.; Pradhan, S. Human maintenance DNA (cytosine-5)-methyltransferase and p53 modulate expression of p53-repressed promoters. *Proc. Natl. Acad. Sci. USA* **2005**, *102*, 1000–1005. [CrossRef]
50. Brenner, C.; Deplus, R.; Didelot, C.; Loriot, A.; Viré, E.; De Smet, C.; Gutierrez, A.; Danovi, D.; Bernard, D.; Boon, T.; et al. Myc represses transcription through recruitment of DNA methyltransferase corepressor. *EMBO J.* **2005**, *24*, 336–346. [CrossRef]
51. Dunican, D.S.; Ruzov, A.; Hackett, J.A.; Meehan, R.R. xDnmt1 regulates transcriptional silencing in pre-MBT *Xenopus* embryos independently of its catalytic function. *Development* **2008**, *135*, 1295–1302. [CrossRef] [PubMed]
52. Clements, E.G.; Mohammad, H.P.; Leadem, B.R.; Easwaran, H.; Cai, Y.; Van Neste, L.; Baylin, S.B. DNMT1 modulates gene expression without its catalytic activity partially through its interactions with histone-modifying enzymes. *Nucleic Acids Res.* **2012**, *40*, 4334–4346. [CrossRef] [PubMed]
53. Kumar, D.; Lassar, A.B. Fibroblast growth factor maintains chondrogenic potential of limb bud mesenchymal cells by modulating DNMT3A recruitment. *Cell Rep.* **2014**, *8*, 1419–1431. [CrossRef] [PubMed]
54. Petell, C.J.; Alabdi, L.; He, M.; San Miguel, P.; Rose, R.; Gowher, H. An epigenetic switch regulates de novo DNA methylation at a subset of pluripotency gene enhancers during embryonic stem cell differentiation. *Nucleic Acids Res.* **2016**, *44*, 7605–7617. [CrossRef] [PubMed]
55. Hnisz, D.; Shrinivas, K.; Young, R.A.; Chakraborty, A.K.; Sharp, P.A. A Phase separation model for transcriptional control. *Cell* **2017**, *169*, 13–23. [CrossRef] [PubMed]
56. Estève, P.-O.; Chin, H.G.; Smallwood, A.; Feehery, G.R.; Gangisetty, O.; Karpf, A.R.; Carey, M.F.; Pradhan, S. Direct interaction between DNMT1 and G9a coordinates DNA and histone methylation during replication. *Genes Dev.* **2006**, *20*, 3089–3103. [CrossRef] [PubMed]
57. Rose, N.R.; Klose, R.J. Understanding the relationship between DNA methylation and histone lysine methylation. *Biochim. Biophys. Acta* **2014**, *1839*, 1362–1372. [CrossRef] [PubMed]
58. Déjardin, J. Switching between epigenetic states at pericentromeric heterochromatin. *Trends Genet. TIG* **2015**, *31*, 661–672. [CrossRef]

59. Li, H.; Rauch, T.; Chen, Z.-X.; Szabó, P.E.; Riggs, A.D.; Pfeifer, G.P. The histone methyltransferase SETDB1 and the DNA methyltransferase DNMT3A interact directly and localize to promoters silenced in cancer cells. *J. Biol. Chem.* **2006**, *281*, 19489–19500. [CrossRef]
60. Rush, M.; Appanah, R.; Lee, S.; Lam, L.L.; Goyal, P.; Lorincz, M.C. Targeting of EZH2 to a defined genomic site is sufficient for recruitment of Dnmt3a but not de novo DNA methylation. *Epigenetics* **2009**, *4*, 404–414. [CrossRef]
61. Chang, Y.; Sun, L.; Kokura, K.; Horton, J.R.; Fukuda, M.; Espejo, A.; Izumi, V.; Koomen, J.M.; Bedford, M.T.; Zhang, X.; et al. MPP8 mediates the interactions between DNA methyltransferase Dnmt3a and H3K9 methyltransferase GLP/G9a. *Nat. Commun.* **2011**, *2*, 533. [CrossRef] [PubMed]
62. Fuks, F.; Burgers, W.A.; Godin, N.; Kasai, M.; Kouzarides, T. Dnmt3a binds deacetylases and is recruited by a sequence-specific repressor to silence transcription. *EMBO J.* **2001**, *20*, 2536–2544. [CrossRef] [PubMed]
63. Borgel, J.; Guibert, S.; Li, Y.; Chiba, H.; Schübeler, D.; Sasaki, H.; Forné, T.; Weber, M. Targets and dynamics of promoter DNA methylation during early mouse development. *Nat. Genet.* **2010**, *42*, 1093–1100. [CrossRef] [PubMed]
64. Velasco, G.; Hubé, F.; Rollin, J.; Neuillet, D.; Philippe, C.; Bouzinba-Segard, H.; Galvani, A.; Viegas-Péquignot, E.; Francastel, C. Dnmt3b recruitment through E2F6 transcriptional repressor mediates germ-line gene silencing in murine somatic tissues. *Proc. Natl. Acad. Sci. USA* **2010**, *107*, 9281–9286. [CrossRef] [PubMed]
65. Auclair, G.; Borgel, J.; Sanz, L.A.; Vallet, J.; Guibert, S.; Dumas, M.; Cavelier, P.; Girardot, M.; Forné, T.; Feil, R.; et al. EHMT2 directs DNA methylation for efficient gene silencing in mouse embryos. *Genome Res.* **2016**, *26*, 192–202. [CrossRef] [PubMed]
66. Endoh, M.; Endo, T.A.; Shinga, J.; Hayashi, K.; Farcas, A.; Ma, K.-W.; Ito, S.; Sharif, J.; Endoh, T.; Onaga, N.; et al. PCGF6-PRC1 suppresses premature differentiation of mouse embryonic stem cells by regulating germ cell-related genes. *eLife* **2017**, *6*. [CrossRef]
67. Mochizuki, K.; Tachibana, M.; Saitou, M.; Tokitake, Y.; Matsui, Y. Implication of DNA demethylation and bivalent histone modification for selective gene regulation in mouse primordial germ cells. *PLoS ONE* **2012**, *7*, e46036. [CrossRef]
68. Quinn, J.J.; Chang, H.Y. Unique features of long non-coding RNA biogenesis and function. *Nat. Rev. Genet.* **2016**, *17*, 47–62. [CrossRef]
69. Schmitz, K.-M.; Mayer, C.; Postepska, A.; Grummt, I. Interaction of noncoding RNA with the rDNA promoter mediates recruitment of DNMT3b and silencing of rRNA genes. *Genes Dev.* **2010**, *24*, 2264–2269. [CrossRef]
70. O'Leary, V.B.; Ovsepian, S.V.; Carrascosa, L.G.; Buske, F.A.; Radulovic, V.; Niyazi, M.; Moertl, S.; Trau, M.; Atkinson, M.J.; Anastasov, N. PARTICLE, a triplex-forming long ncRNA, regulates locus-specific methylation in response to low-dose irradiation. *Cell Rep.* **2015**, *11*, 474–485. [CrossRef]
71. O'Leary, V.B.; Hain, S.; Maugg, D.; Smida, J.; Azimzadeh, O.; Tapio, S.; Ovsepian, S.V.; Atkinson, M.J. Long non-coding RNA PARTICLE bridges histone and DNA methylation. *Sci. Rep.* **2017**, *7*, 1790. [CrossRef]
72. Barlow, D.P.; Bartolomei, M.S. Genomic imprinting in mammals. *Cold Spring Harb. Perspect. Biol.* **2014**, *6*. [CrossRef] [PubMed]
73. Mohammad, F.; Mondal, T.; Guseva, N.; Pandey, G.K.; Kanduri, C. *Kcnq1ot1* noncoding RNA mediates transcriptional gene silencing by interacting with Dnmt1. *Development* **2010**, *137*, 2493–2499. [CrossRef] [PubMed]
74. Chalei, V.; Sansom, S.N.; Kong, L.; Lee, S.; Montiel, J.F.; Vance, K.W.; Ponting, C.P. The long non-coding RNA *Dali* is an epigenetic regulator of neural differentiation. *eLife* **2014**, *3*, e04530. [CrossRef]
75. Johnsson, P.; Ackley, A.; Vidarsdottir, L.; Lui, W.-O.; Corcoran, M.; Grandér, D.; Morris, K.V. A pseudogene long-noncoding-RNA network regulates PTEN transcription and translation in human cells. *Nat. Struct. Mol. Biol.* **2013**, *20*, 440–446. [CrossRef] [PubMed]
76. Stathopoulou, A.; Chhetri, J.B.; Ambrose, J.C.; Estève, P.-O.; Ji, L.; Erdjument-Bromage, H.; Zhang, G.; Neubert, T.A.; Pradhan, S.; Herrero, J.; et al. A novel requirement for DROSHA in maintenance of mammalian CG methylation. *Nucleic Acids Res.* **2017**, *45*, 9398–9412. [CrossRef]
77. Zhang, G.; Estève, P.-O.; Chin, H.G.; Terragni, J.; Dai, N.; Corrêa, I.R.; Pradhan, S. Small RNA-mediated DNA (cytosine-5) methyltransferase 1 inhibition leads to aberrant DNA methylation. *Nucleic Acids Res.* **2015**, *43*, 6112–6124. [CrossRef]

78. Thompson, P.J.; Macfarlan, T.S.; Lorincz, M.C. Long terminal repeats: From parasitic elements to building blocks of the transcriptional regulatory repertoire. *Mol. Cell* **2016**, *62*, 766–776. [CrossRef]
79. Chuong, E.B.; Elde, N.C.; Feschotte, C. Regulatory activities of transposable elements: From conflicts to benefits. *Nat. Rev. Genet.* **2017**, *18*, 71–86. [CrossRef]
80. Rodriguez-Terrones, D.; Torres-Padilla, M.-E. Nimble and ready to mingle: Transposon outbursts of early development. *Trends Genet. TIG* **2018**, *34*, 806–820. [CrossRef]
81. Ernst, C.; Odom, D.T.; Kutter, C. The emergence of piRNAs against transposon invasion to preserve mammalian genome integrity. *Nat. Commun.* **2017**, *8*, 1411. [CrossRef] [PubMed]
82. Turelli, P.; Castro-Diaz, N.; Marzetta, F.; Kapopoulou, A.; Raclot, C.; Duc, J.; Tieng, V.; Quenneville, S.; Trono, D. Interplay of TRIM28 and DNA methylation in controlling human endogenous retroelements. *Genome Res.* **2014**, *24*, 1260–1270. [CrossRef] [PubMed]
83. Kuramochi-Miyagawa, S.; Watanabe, T.; Gotoh, K.; Totoki, Y.; Toyoda, A.; Ikawa, M.; Asada, N.; Kojima, K.; Yamaguchi, Y.; Ijiri, T.W.; et al. DNA methylation of retrotransposon genes is regulated by Piwi family members MILI and MIWI2 in murine fetal testes. *Genes Dev.* **2008**, *22*, 908–917. [CrossRef] [PubMed]
84. Aravin, A.A.; Sachidanandam, R.; Bourc'his, D.; Schaefer, C.; Pezic, D.; Toth, K.F.; Bestor, T.; Hannon, G.J. A piRNA pathway primed by individual transposons is linked to de novo DNA methylation in mice. *Mol. Cell* **2008**, *31*, 785–799. [CrossRef] [PubMed]
85. Itou, D.; Shiromoto, Y.; Yukiho, S.; Ishii, C.; Nishimura, T.; Ogonuki, N.; Ogura, A.; Hasuwa, H.; Fujihara, Y.; Kuramochi-Miyagawa, S.; et al. Induction of DNA methylation by artificial piRNA production in male germ cells. *Curr. Biol. CB* **2015**, *25*, 901–906. [CrossRef] [PubMed]
86. Watanabe, T.; Tomizawa, S.; Mitsuya, K.; Totoki, Y.; Yamamoto, Y.; Kuramochi-Miyagawa, S.; Iida, N.; Hoki, Y.; Murphy, P.J.; Toyoda, A.; et al. Role for piRNAs and noncoding RNA in de novo DNA methylation of the imprinted mouse *Rasgrf1* locus. *Science* **2011**, *332*, 848–852. [CrossRef]
87. Liu, J.; Zhang, S.; Cheng, B. Epigenetic roles of PIWI-interacting RNAs (piRNAs) in cancer metastasis (Review). *Oncol. Rep.* **2018**, *40*, 2423–2434. [CrossRef]
88. Leonhardt, H.; Page, A.W.; Weier, H.U.; Bestor, T.H. A targeting sequence directs DNA methyltransferase to sites of DNA replication in mammalian nuclei. *Cell* **1992**, *71*, 865–873. [CrossRef]
89. Iida, T.; Suetake, I.; Tajima, S.; Morioka, H.; Ohta, S.; Obuse, C.; Tsurimoto, T. PCNA clamp facilitates action of DNA cytosine methyltransferase 1 on hemimethylated DNA. *Genes Cells* **2002**, *7*, 997–1007. [CrossRef]
90. Boehm, E.M.; Washington, M.T. R.I.P. to the PIP: PCNA-binding motif no longer considered specific: PIP motifs and other related sequences are not distinct entities and can bind multiple proteins involved in genome maintenance. *BioEssays* **2016**, *38*, 1117–1122. [CrossRef]
91. Schermelleh, L.; Haemmer, A.; Spada, F.; Rösing, N.; Meilinger, D.; Rothbauer, U.; Cardoso, M.C.; Leonhardt, H. Dynamics of Dnmt1 interaction with the replication machinery and its role in postreplicative maintenance of DNA methylation. *Nucleic Acids Res.* **2007**, *35*, 4301–4312. [CrossRef] [PubMed]
92. Sharif, J.; Muto, M.; Takebayashi, S.; Suetake, I.; Iwamatsu, A.; Endo, T.A.; Shinga, J.; Mizutani-Koseki, Y.; Toyoda, T.; Okamura, K.; et al. The SRA protein Np95 mediates epigenetic inheritance by recruiting Dnmt1 to methylated DNA. *Nature* **2007**, *450*, 908–912. [CrossRef] [PubMed]
93. Bostick, M.; Kim, J.K.; Estève, P.-O.; Clark, A.; Pradhan, S.; Jacobsen, S.E. UHRF1 plays a role in maintaining DNA methylation in mammalian cells. *Science* **2007**, *317*, 1760–1764. [CrossRef] [PubMed]
94. Achour, M.; Jacq, X.; Rondé, P.; Alhosin, M.; Charlot, C.; Chataigneau, T.; Jeanblanc, M.; Macaluso, M.; Giordano, A.; Hughes, A.D.; et al. The interaction of the SRA domain of ICBP90 with a novel domain of DNMT1 is involved in the regulation of *VEGF* gene expression. *Oncogene* **2008**, *27*, 2187–2197. [CrossRef] [PubMed]
95. Bashtrykov, P.; Jankevicius, G.; Jurkowska, R.Z.; Ragozin, S.; Jeltsch, A. The UHRF1 protein stimulates the activity and specificity of the maintenance DNA methyltransferase DNMT1 by an allosteric mechanism. *J. Biol. Chem.* **2014**, *289*, 4106–4115. [CrossRef] [PubMed]
96. Berkyurek, A.C.; Suetake, I.; Arita, K.; Takeshita, K.; Nakagawa, A.; Shirakawa, M.; Tajima, S. The DNA methyltransferase Dnmt1 directly interacts with the SET and RING finger-associated (SRA) domain of the multifunctional protein Uhrf1 to facilitate accession of the catalytic center to hemi-methylated DNA. *J. Biol. Chem.* **2014**, *289*, 379–386. [CrossRef] [PubMed]

97. Smets, M.; Link, S.; Wolf, P.; Schneider, K.; Solis, V.; Ryan, J.; Meilinger, D.; Qin, W.; Leonhardt, H. DNMT1 mutations found in HSANIE patients affect interaction with UHRF1 and neuronal differentiation. *Hum. Mol. Genet.* **2017**, *26*, 1522–1534. [CrossRef] [PubMed]
98. Ishiyama, S.; Nishiyama, A.; Saeki, Y.; Moritsugu, K.; Morimoto, D.; Yamaguchi, L.; Arai, N.; Matsumura, R.; Kawakami, T.; Mishima, Y.; et al. Structure of the Dnmt1 reader module complexed with a unique two-mono-ubiquitin mark on histone H3 reveals the basis for DNA methylation maintenance. *Mol. Cell* **2017**, *68*, 350–360.e7. [CrossRef]
99. Xu, C.; Corces, V.G. Nascent DNA methylome mapping reveals inheritance of hemimethylation at CTCF/cohesin sites. *Science* **2018**, *359*, 1166–1170. [CrossRef]
100. Charlton, J.; Downing, T.L.; Smith, Z.D.; Gu, H.; Clement, K.; Pop, R.; Akopian, V.; Klages, S.; Santos, D.P.; Tsankov, A.M.; et al. Global delay in nascent strand DNA methylation. *Nat. Struct. Mol. Biol.* **2018**, *25*, 327–332. [CrossRef]
101. Ferry, L.; Fournier, A.; Tsusaka, T.; Adelmant, G.; Shimazu, T.; Matano, S.; Kirsh, O.; Amouroux, R.; Dohmae, N.; Suzuki, T.; et al. Methylation of DNA ligase 1 by G9a/GLP recruits UHRF1 to replicating DNA and regulates DNA methylation. *Mol. Cell* **2017**, *67*, 550–565.e5. [CrossRef] [PubMed]
102. Zhou, T.; Xiong, J.; Wang, M.; Yang, N.; Wong, J.; Zhu, B.; Xu, R.-M. Structural basis for hydroxymethylcytosine recognition by the SRA domain of UHRF2. *Mol. Cell* **2014**, *54*, 879–886. [CrossRef] [PubMed]
103. Mortusewicz, O.; Schermelleh, L.; Walter, J.; Cardoso, M.C.; Leonhardt, H. Recruitment of DNA methyltransferase I to DNA repair sites. *Proc. Natl. Acad. Sci. USA* **2005**, *102*, 8905–8909. [CrossRef] [PubMed]
104. Ruzov, A.; Shorning, B.; Mortusewicz, O.; Dunican, D.S.; Leonhardt, H.; Meehan, R.R. MBD4 and MLH1 are required for apoptotic induction in xDNMT1-depleted embryos. *Development* **2009**, *136*, 2277–2286. [CrossRef] [PubMed]
105. Cuozzo, C.; Porcellini, A.; Angrisano, T.; Morano, A.; Lee, B.; Di Pardo, A.; Messina, S.; Iuliano, R.; Fusco, A.; Santillo, M.R.; et al. DNA damage, homology-directed repair, and DNA methylation. *PLoS Genet.* **2007**, *3*, e110. [CrossRef]
106. Morano, A.; Angrisano, T.; Russo, G.; Landi, R.; Pezone, A.; Bartollino, S.; Zuchegna, C.; Babbio, F.; Bonapace, I.M.; Allen, B.; et al. Targeted DNA methylation by homology-directed repair in mammalian cells. Transcription reshapes methylation on the repaired gene. *Nucleic Acids Res.* **2014**, *42*, 804–821. [CrossRef]
107. Ha, K.; Lee, G.E.; Palii, S.S.; Brown, K.D.; Takeda, Y.; Liu, K.; Bhalla, K.N.; Robertson, K.D. Rapid and transient recruitment of DNMT1 to DNA double-strand breaks is mediated by its interaction with multiple components of the DNA damage response machinery. *Hum. Mol. Genet.* **2011**, *20*, 126–140. [CrossRef]
108. Tian, Y.; Paramasivam, M.; Ghosal, G.; Chen, D.; Shen, X.; Huang, Y.; Akhter, S.; Legerski, R.; Chen, J.; Seidman, M.M.; et al. UHRF1 contributes to DNA damage repair as a lesion recognition factor and nuclease scaffold. *Cell Rep.* **2015**, *10*, 1957–1966. [CrossRef]
109. Liang, C.-C.; Zhan, B.; Yoshikawa, Y.; Haas, W.; Gygi, S.P.; Cohn, M.A. UHRF1 is a sensor for DNA interstrand crosslinks and recruits FANCD2 to initiate the Fanconi anemia pathway. *Cell Rep.* **2015**, *10*, 1947–1956. [CrossRef]
110. Miotto, B.; Chibi, M.; Xie, P.; Koundrioukoff, S.; Moolman-Smook, H.; Pugh, D.; Debatisse, M.; He, F.; Zhang, L.; Defossez, P.-A. The RBBP6/ZBTB38/MCM10 axis regulates DNA replication and common fragile site stability. *Cell Rep.* **2014**, *7*, 575–587. [CrossRef]
111. Van Houten, B.; Santa-Gonzalez, G.A.; Camargo, M. DNA repair after oxidative stress: Current challenges. *Curr. Opin. Toxicol.* **2018**, *7*, 9–16. [CrossRef] [PubMed]
112. O'Hagan, H.M.; Wang, W.; Sen, S.; Destefano Shields, C.; Lee, S.S.; Zhang, Y.W.; Clements, E.G.; Cai, Y.; Van Neste, L.; Easwaran, H.; et al. Oxidative damage targets complexes containing DNA methyltransferases, SIRT1, and polycomb members to promoter CpG Islands. *Cancer Cell* **2011**, *20*, 606–619. [CrossRef] [PubMed]
113. Ding, N.; Bonham, E.M.; Hannon, B.E.; Amick, T.R.; Baylin, S.B.; O'Hagan, H.M. Mismatch repair proteins recruit DNA methyltransferase 1 to sites of oxidative DNA damage. *J. Mol. Cell Biol.* **2016**, *8*, 244–254. [CrossRef] [PubMed]
114. Maiuri, A.R.; Peng, M.; Podicheti, R.; Sriramkumar, S.; Kamplain, C.M.; Rusch, D.B.; DeStefano Shields, C.E.; Sears, C.L.; O'Hagan, H.M. Mismatch repair proteins initiate epigenetic alterations during inflammation-driven tumorigenesis. *Cancer Res.* **2017**, *77*, 3467–3478. [CrossRef] [PubMed]

115. Gu, T.; Lin, X.; Cullen, S.M.; Luo, M.; Jeong, M.; Estecio, M.; Shen, J.; Hardikar, S.; Sun, D.; Su, J.; et al. DNMT3A and TET1 cooperate to regulate promoter epigenetic landscapes in mouse embryonic stem cells. *Genome Biol.* **2018**, *19*, 88. [CrossRef] [PubMed]
116. Noh, K.-M.; Wang, H.; Kim, H.R.; Wenderski, W.; Fang, F.; Li, C.H.; Dewell, S.; Hughes, S.H.; Melnick, A.M.; Patel, D.J.; et al. Engineering of a histone-recognition domain in Dnmt3a alters the epigenetic landscape and phenotypic features of mouse ESCs. *Mol. Cell* **2015**, *59*, 89–103. [CrossRef] [PubMed]
117. Holtzman, L.; Gersbach, C.A. Editing the epigenome: Reshaping the genomic landscape. *Annu. Rev. Genomics Hum. Genet.* **2018**, *19*, 43–71. [CrossRef]
118. Di Ruscio, A.; Ebralidze, A.K.; Benoukraf, T.; Amabile, G.; Goff, L.A.; Terragni, J.; Figueroa, M.E.; De Figueiredo Pontes, L.L.; Alberich-Jorda, M.; Zhang, P.; et al. DNMT1-interacting RNAs block gene-specific DNA methylation. *Nature* **2013**, *503*, 371–376. [CrossRef]
119. Pal, S.; Tyler, J.K. Epigenetics and aging. *Sci. Adv.* **2016**, *2*, e1600584. [CrossRef]

© 2018 by the authors. Licensee MDPI, Basel, Switzerland. This article is an open access article distributed under the terms and conditions of the Creative Commons Attribution (CC BY) license (http://creativecommons.org/licenses/by/4.0/).

Review

The Growing Complexity of UHRF1-Mediated Maintenance DNA Methylation

Si Xie and Chengmin Qian *

School of Biomedical Sciences, The University of Hong Kong, Hong Kong, China; h1094160@connect.hku.hk
* Correspondence: cmqian@hku.hk; Tel.: +0852-39-176-820.

Received: 1 November 2018; Accepted: 29 November 2018; Published: 3 December 2018

Abstract: Mammalian DNMT1 is mainly responsible for maintenance DNA methylation that is critical in maintaining stem cell pluripotency and controlling lineage specification during early embryonic development. A number of studies have demonstrated that DNMT1 is an auto-inhibited enzyme and its enzymatic activity is allosterically regulated by a number of interacting partners. UHRF1 has previously been reported to regulate DNMT1 in multiple ways, including control of substrate specificity and the proper genome targeting. In this review, we discuss the recent advances in our understanding of the regulation of DNMT1 enzymatic activity by UHRF1 and highlight a number of unresolved questions.

Keywords: DNA methylation; DNMT1; UHRF1; USP7; ubiquitination

1. Introduction

DNA methylation is one of the best-characterized epigenetic changes that have a critical role in numerous biological processes including gene expression, genomic imprinting, X chromosome inactivation and genome stability [1–3]. In mammals, DNA methylation predominantly occurs at the C5 position of cytosine in CpG dinucleotides [4,5]. DNA methylation is catalyzed by the DNA methyltransferase family members including DNMT1, DNMT3A and DNMT3B. Another enzymatically inactive member DNMT3L interacts with both DNMT3A and DNMT3B and stimulates their enzymatic activity.

Methylation on CpG sites in promoters generally leads to gene repression by either inhibiting transcription factor binding or recruiting the repressive complex to the promoter region [6]. Therefore, selective methylation on promoter regions of certain genes facilitates the establishment of specific gene expression pattern in differentiated cells. The methylation pattern is tissue-specific and built mainly by de novo DNA methyltransferases DNMT3A and DNMT3B [7–9]. Once established, maintenance DNA methyltransferase DNMT1 ensures that the methylation pattern is faithfully propagated throughout successive cell divisions to maintain cell-specific functions in differentiated cells. During the replication process, DNMT1 preferentially catalyzes the conversion of hemi-methylated CpG dinucleotides in daughter strands to fully methylated forms. It is worthy to note that DNMT3A and DNMT3B have been found to be involved in maintaining DNA methylation patterns in specific loci, whereas DNMT1 exhibits de novo DNA methylation under certain circumstances [10–12]. In addition, the DNA methylation pattern is dynamic during the developmental processes and to maintain the balance between DNA methylation and demethylation could have a great impact on human health and disease. Similar to other epigenetic modifications, DNA methylation is reversible. 5-Methyl cytosine (5mC) can be further converted to 5-hydroxymethyl cytosine (5hmC), 5-formyl cytosine (5fC) and 5-carboxyl cytosine (5caC) by the DNA methyl-cytosine dioxygenases TET1, TET2 and TET3. It should be pointed out that 5hmC, 5fC and 5caC not just serve as DNA demethylation intermediates but also have distinct regulatory functions in various developmental processes [13,14].

Mounting evidence has demonstrated that DNMT1 alone is not sufficient to maintain the global DNA methylation throughout cell division. Indeed, structural and biochemical analyses have suggested that DNMT1 is a self-inhibited enzyme, as DNMT1 N-terminal regulatory domains including replication foci targeting sequence (RFTS) domain and CXXC domain show autoinhibitory effect on DNMT1 enzymatic activity. Therefore, additional protein factors are required to release DNMT1 from its self-inhibited state. UHRF1 has emerged in recent studies to be one of such regulators [15–17]. In this review, we summarize the most recent findings on UHRF1-mediated regulation of DNMT1 recruitment and activation.

2. DNMT1 is an Autoinhibited DNA Methyltransferase

DNMT1 is a multi-modular protein consisting of a DMAP1-binding domain, RFTS domain, a CXXC domain, two BAH domains and a C-terminal catalytic domain (Figure 1A). Numerous studies have shown that N-terminal domains of DNMT1 have key regulatory functions. For example, the DMAP1-binding domain and RFTS domain together with a PCNA-interacting protein (PIP) box inserted between them are mainly responsible for DNMT1 stability and its proper localization onto DNA replication site [18–20]. At the same time, DNMT1 RFTS domain and CXXC domain can regulate the activity of the catalytic domain. The CXXC domain specifically binds to unmethylated CpG sequences in DNA and positions the CXXC-BAH1 linker between DNA and the catalytic pocket to prevent de novo methylation on those CpG dinucleotides [21,22].

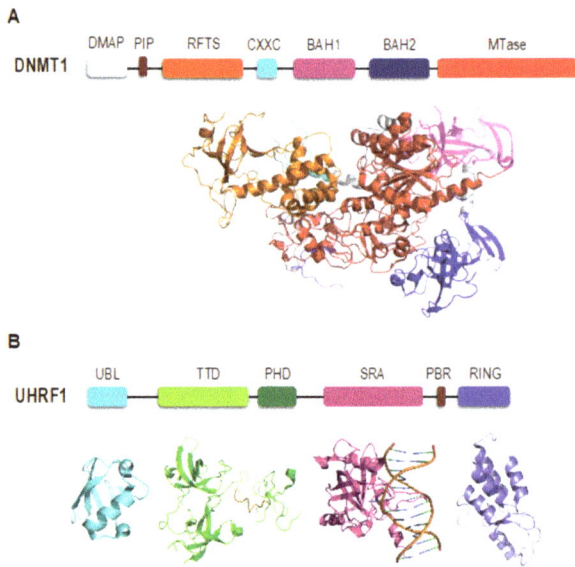

Figure 1. Structures of DNMT1 and UHRF1. (**A**) Domain architecture of DNMT1 and crystal structure of human DNMT1 (aa: 351-1600; PDB: 4WXX) [23]. (**B**) Domain organization of UHRF1 and crystal structures of mouse UHRF1 UBL domain (PDB: 2FAZ), TTD-PHD in complex with H3K9me3 peptide (PDB: 3ASK), SRA in complex with hemi-methylated DNA fragment (PDB: 3CLZ) and UHRF1 RING finger (PDB: 3FL2) [24,25].

Earlier studies have revealed that RFTS domain regulates DNMT1 enzymatic activity through an auto-inhibitory mechanism, as explained by structural analysis showing that RFTS domain occupies DNMT1 catalytic pocket and prevents substrate DNA binding (Figure 2) [23,26]. Consistent with the structural observation that RFTS domain has intramolecular interaction with the catalytic domain (CD), disruption of RFTS-CD interaction strengthens the RFTS binding to histone H3, which led to abnormal DNMT1 accumulation on chromatin during S-phase [27] and aberrant increase on DNMT1 activity both in vitro and in vivo [28,29]. Such fine-tuned regulatory mechanism of DNMT1 methyltransferase activity could guarantee the faithful inheritance of DNA methylation pattern and maintain tissue-specific functions throughout cell division. Therefore, RFTS has been proposed as a fail-safe lock which protects the genome from aberrant replication-independent DNA methylation [30].

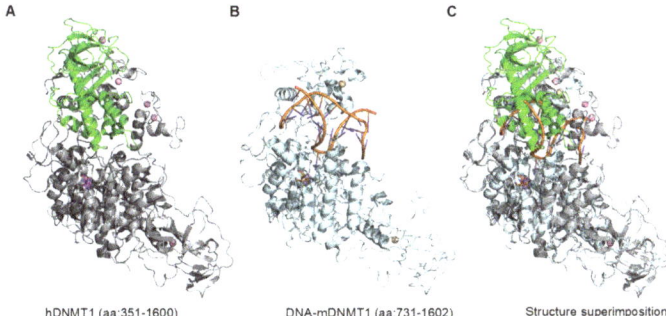

Figure 2. The DNMT1 replication foci targeting sequence (RFTS) domain regulates DNMT1 enzymatic activity through an auto-inhibitory mechanism. (**A**) Crystal structure of human DNMT1 (aa: 351-1600; PDB: 4WXX) in free form [23]. (**B**) Crystal structure of mouse DNMT1 (aa:731-1602; PDB: 4DA4) bound by hemimethylated CpG DNA [31]. (**C**) Superimposition of structures in (**A**) and (**B**) revealed that the RFTS domain occupies DNMT1's catalytic pocket and prevents substrate DNA binding.

3. UHRF1 is Required for Proper Loading of DNMT1 onto Chromatin

UHRF1 is a multifunctional epigenetic regulator that bridges DNA methylation with multiple histone post-translational modifications such as histone methylation, ubiquitination and acetylation. A number of studies have demonstrated that UHRF1 interacts directly with DNMT1 and is essential in DNA methylation maintenance (Figure 1B). The UHRF1 SRA domain specifically binds to hemi-methylated DNA [15,16,25,32–34], that helps direct DNMT1 to natural substrate sites. It has also been reported that UHRF1 SRA domain interacts with DNMT1 RFTS domain which could promote the access of DNMT1's catalytic center to hemi-methylated DNA [29,35]. The UHRF1 PHD finger and tandem tudor domain (TTD) cooperatively recognize H3R2 and H3K9me2/3 mark that may facilitate the proper localization of DNMT1 on genomic loci [24,36–43]. The UHRF1 C-terminal RING finger domain has been shown to function as an E3 ubiquitin ligase responsible for histone H3 ubiquitination [44,45]. Ubiquitinated histone H3 is subsequently recognized by DNMT1 that promotes recruitment of DNMT1 to DNA replication sites [46,47].

4. Allosteric Regulation of UHRF1 E3 Ubiquitin Ligase Activity

UHRF1 is proposed to adopt a closed autoinhibited conformation due to extensive inter-domain interactions within the protein [48–50]. Getalo et al. reported that a polybasic region (PBR) present between SRA and RING finger of UHRF1 serves as a competitive inhibitor of H3K9me2/3 interaction with the UHRF1 TTD, while the binding of phosphatidylinostiol phosphate (PI5P) to PBR allosterically relieves the inhibition [48]. More recent studies have demonstrated that PBR is a versatile platform that interacts with other partners such as USP7 and DNMT1 [49,51,52]. UHRF1 TTD interacts directly with a histone H3K9-like mimic within DNA ligase 1 (LIG1) [53]. All these interactions can similarly trigger UHRF1 conformational changes and regulate intramolecular activation of UHRF1. On the other hand, the binding of hemi-methylated DNA also causes conformational changes in UHRF1 [49], potentially releases TTD and allosterically regulates UHRF1 E3 ubiquitin ligase activity [50]. Multivalent engagement of hemi-methylated linker DNA and H3K9me2 enhances the enzymatic activity of UHRF1 toward nucleosomal histone substrates [54].

Moreover, UHRF1 UBL domain has just been reported to allosterically regulate UHRF1 E3 ligase activity towards histone H3 [55,56]. UHRF1 UBL and RING finger interact with the ubiquitin conjugating E2 enzyme UbcH5a to stimulate histone H3 ubiquitination.

5. UHRF1-Dependent Histone H3 Ubiquitination Stimulates DNMT1 Enzymatic Activity

Nishiyama et al. first demonstrated that UHRF1 can ubiquitinate histone H3 on K23, and ubiquitinated H3 is subsequently recognized by DNMT1 [46]. The process actually facilitates the recruitment of DNMT1 to the replication foci. Several studies have further shown that a number of lysine residues including K14, K18 and K23 on histone H3 can be ubiquitinated by UHRF1 [47,50]. Two recent structure-based studies have reported crystal structures of RFTS in complex with ubiquitin and ubiquitinated histone H3 (H3-K18Ub/K23Ub) respectively, which shed light on the molecular mechanism of how UHRF1-dependent H3 ubiquitination regulates DNA methylation maintenance [57,58]. Notably, the RFTS domain of DNMT1 binds ubiquitin with 1:2 stoichiometry in both structures (Figure 3A,B). Superimposition of two complex structures gives a RMSD of 0.78 Å (Figure 3C), consistent with almost the same RFTS/ubiquitin interaction network observed in both structures. Although the ubiquitin-conjugated histone H3 is directly involved in RFTS interaction, it has no obvious effect on ubiquitin recognition by RFTS. This is consistent with the notion in the field that ubiquitin-conjugated substrates usually have no visible impact on the interaction between ubiquitin and ubiquitin-binding domains. RFTS binds to two mono-ubiquitinated histone H3 (H3Ub2) approximately 100 times stronger than histone H3 alone. Intriguingly, RFTS exhibits similar binding affinities to single mono-ubiquitinated histone H3 (H3Ub) and the unmodified H3, suggesting that binding of RFTS to histone H3 and two covalently attached ubiquitin molecules is cooperative [57]. Consistently, mutations on RFTS residues that only participate in the interaction with one of the bound ubiquitin molecules, completely abolish the RFTS association with both ubiquitin molecules, implying that binding of two ubiquitin molecules is interdependent and single mono-ubiquitinated histone H3 is insufficient for DNMT1 recruitment [57,58]. More interestingly, in vitro DNA methylation assays also demonstrate that only two mono-ubiquitinated histone H3, but not ubiquitin alone or unmodified histone H3 could stimulate DNMT1 enzymatic activity [57,58]. However, currently, available structures are unable to explain how exactly the dual mono-ubiquitinated histone H3 alleviates the RFTS auto-inhibitory effect. Further efforts on structural analysis of the complex of two mono-ubiquitinated histone H3 with DNMT1 including RFTS and catalytic domain are required to demonstrate the detailed conformational changes in DNMT1 upon ubiquitinated histone H3 binding.

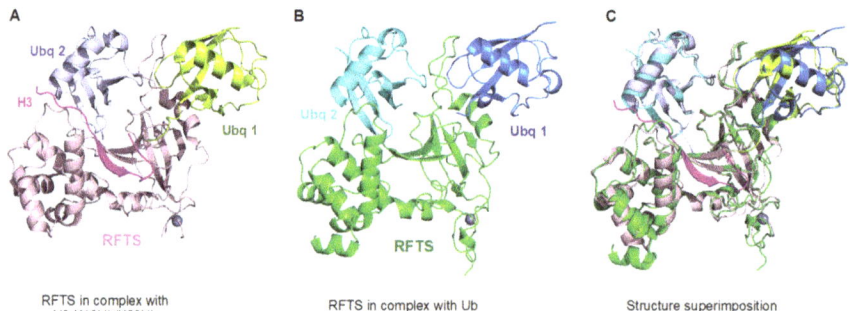

Figure 3. The DNMT1 RFTS domain is a dual mono-ubiquitinated histone H3 binding module. (**A**) Crystal structure of DNMT1 RFTS in complex with ubiquitinated histone H3 (PDB: 5WVO) [57]. (**B**) Crystal structure of DNMT1 RFTS in complex with ubiquitin (PDB: 5YDR) [58]. (**C**) The interaction between RFTS and ubiquitin is essentially the same in these structures. Superimposition of two complex structures gives an RMSD of 0.78Å.

6. DNMT1 Recruitment and Stimulation by UHRF1 N-Terminal UBL Domain

Considering that UHRF1 N-terminal UBL domain structurally resembles ubiquitin, Li et al. went on to test if the UBL domain of UHRF1 could interact with DNMT1 RFTS domain [58]. Indeed, UHRF1 UBL was found to physically interact with DNMT1 RFTS domain. Strikingly, the interaction network of RFTS-UBL is distinct from that of RFTS-ubiquitin although UBL shares high structural similarity with ubiquitin. For example, mutations on RFTS residues that abolished ubiquitin binding showed little or no effect on the interaction with UBL. Unexpectedly, the authors also found that interaction between DNMT1 and UHRF1 UBL could stimulate DNMT1 enzymatic activity. UHRF1 UBL could stimulate the enzymatic activity of DNMT1 segments (aa: 621-1616) and (aa: 351-1616) to a similar level, but H3Ub2 shows no stimulatory effect toward DNMT1(aa: 621-1616), suggesting distinct stimulatory mechanisms by UHRF1 UBL domain and ubiquitinated histone H3 [58]. The authors were able to figure out that UHRF1 UBL not only physically interacts with RFTS but also binds to DNMT1 (aa: 621-1616) and the binding to this DNMT1 fragment is essential for its stimulatory effect on DNMT1 enzymatic activity. It is worthwhile elucidating the detailed molecular basis of the interaction between UHRF1 UBL and DNMT1 in future studies.

UHRF1 UBL has also been revealed to interact with the ubiquitin conjugating E2 enzyme UbcH5a to stabilize the active E2/UHRF1/ubiquitin complex and direct mono-ubiquitination towards histone H3 [55,56], which eventually facilitates the recruitment of DNMT1 to replication foci for efficient DNA methylation maintenance.

7. Epigenetic Crosstalk between DNA Methylation and Histone Modifications in Maintenance DNA Methylation

Mounting evidence suggests that crosstalk between DNA methylation and histone modifications occurs to maintain DNA methylation patterns during replication. Previous reports have indicated that binding of UHRF1 to H3K9me2/3 is indispensable for DNMT1 association with chromatin. Disruption of the interaction between UHRF1 and H3K9me2/3 or H3R2 leads to loss of DNMT1 chromatin localization and global hypomethylation [36,41,59]. In contrast, a recent study demonstrated that although disruption of UHRF1 and H3K9me2/3 association leads to hypomethylation, the reduction of DNA methylation occurs globally and does not restrict in the H3K9me2/3 enriched region, suggesting that DNMT1 mediating DNA maintenance methylation is largely independent of H3K9 methylation [59]. Actually, this observation is not surprising considering that the UHRF1 PHD finger but not TTD plays a dominant role in histone H3 binding [24,40,43,60], and expression of UHRF1 with a defective PHD finger failed to restore DNA methylation pattern in UHRF1 knockout cells [47]. The UHRF1 PHD finger recognizes the histone H3 N-terminal tail with unmodified R2. Consequently, methylation of H3R2 by PRMT6 impairs chromatin binding of UHRF1 and induces global DNA hypomethylation [61]. Interestingly, a recent report suggested that side chain guanidinium group of histone H3R8 forms one hydrogen bond with the carbonyl oxygen atom of DNMT1 Tyr564, and forms another hydrogen bond with the side chain of DNMT1 Glu572 [57]. Two previous studies showed that side chain NηH atoms of histone H3R8 form hydrogen bonds with the carbonyl oxygen atom of UHRF1 Asp190 [24,60]. It is likely that methylation on histone H3R8 will antagonize its binding to DNMT1 and UHRF1 based on the above three structure-based analyses.

Histone H3 N-terminal residues such as K14, K18 and K23 can be acetylated or ubiquitinated, acetylation of these residues would preclude ubiquitination on the same sites. Previous studies have shown that HDAC1/2 associate with nascent chromatin during replication [62], DNMT1 and HDAC1/2 reside in a large protein complex [19,63], the presence of HDACs likely makes H3K14, H3K18 and H3K23 sites accessible to UHRF1 for mono-ubiquitination at replicating chromatin of specific genomic loci. On the other hand, acetylation on these lysine residues occurs more frequently at transcriptionally active regions where DNA methylation is rare. Furthermore, a number of studies have shown that DNMT1 and UHRF1 are subject to acetylation, ubiquitination and other post-translational modifications. These post-translational modifications (PTMs) could have an impact on protein stability, modulating protein-protein interactions and even their enzymatic activities, adding another layer of complexity to the regulatory mechanism of maintenance DNA methylation. Future studies are needed to examine if these epigenetic marks are sequentially erased or established during maintenance DNA methylation.

8. A Working Model for DNMT1/UHRF1 Complex in Maintenance DNA Methylation

Based on the current knowledge, we propose a simplified working model for the DNMT1/UHRF1 complex in maintenance DNA methylation (Figure 4): (i) The UHRF1 TTD interacts with PBR motif, and the SRA domain binds to the PHD domain. These intramolecular interactions lock UHRF1 in a closed conformation and keep it in an inactive state [48–52]. (ii) During replication, the UHRF1 SRA domain preferentially binds to hemi-methylated DNA, which triggers a conformational change in UHRF1 to dissociate the PHD finger from SRA and release TTD from the PBR motif, allowing TTD-PHD to interact with unmodified H3R2 and H3K9me2/3 marks. This process could be facilitated by the binding of USP7 or PIP5 to PBR. Multivalent engagement of chromatin modifications by various modules of UHRF1 would shift its conformation into an open state in which UHRF1's E3 ubiquitin ligase activity is greatly stimulated, hence, UHRF1 can effectively add mono-ubiquitin to multiple lysine residues of histone H3. (iii) Through the physical interaction with a number of partner proteins such as PCNA, UHRF1 and ubiquitinated histone H3, DNMT1 is loaded onto chromatin and ready to catalyze methylation on hemi-methylated DNA. The UHRF1-dependent ubiquitination of histone H3 dramatically enhances its interaction with the DNMT1 RFTS domain and displaces the TTD-PHD module of UHRF1 from histone H3. Binding of dual mono-ubiquitinated H3 to DNMT1 will disrupt the intramolecular interaction of RFTS with the C-terminal catalytic region and allow the hemi-methylated DNA substrate to access the catalytic center. (iv) After a hemi-methylated CpG is converted into the fully methylated state, DNMT1 dissociates from the ubiquitinated histone H3 and moves along newly replicated DNA to carry out processive methylation reactions. Again, UHRF1 binds to hemi-methylated DNA first and adds ubiquitin marks to neighboring histone H3. At the same time, the deubiquitinase USP7 is activated to erase ubiquitin moieties from histone H3 at methyl DNA recovered regions [64]. Likely de-ubiquitination promotes the release of DNMT1 from histone H3 to ensure processive methylation reaction to proceed.

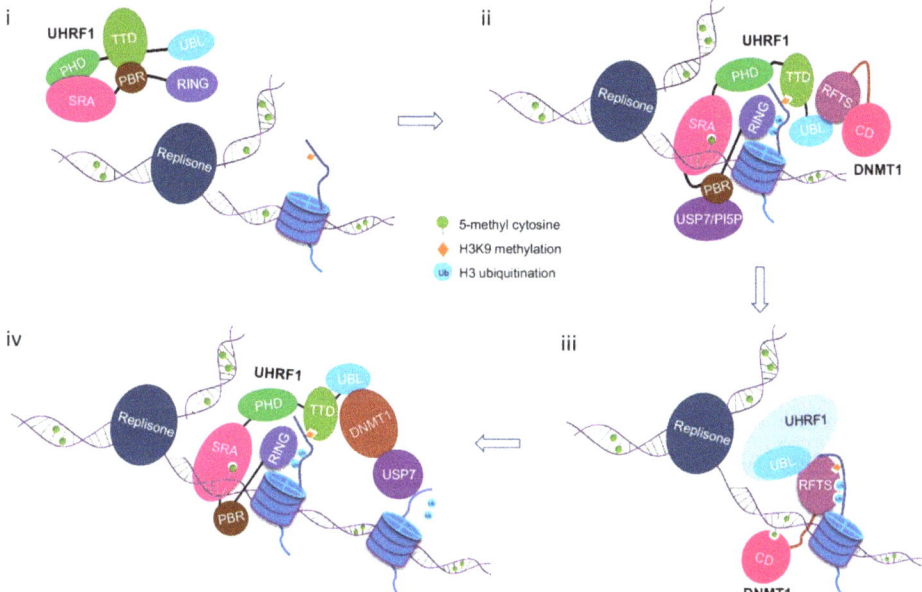

Figure 4. A simplified model for the function of DNMT1/UHRF1 complex in maintenance DNA methylation. (i) UHRF1 adopts a closed conformation in which TTD binds to the PBR motif between SRA and RING finger. (ii) A conformational switch in UHRF1 from inactive closed conformation to active open state is triggered by hemi-methylated DNA binding (and may be facilitated by USP7 or PI5P binding to PBR), which allows TTD-PHD bind to H3K9me2/3 and the RING finger to add ubiquitin to multiple lysine residues of histone H3. (iii) Binding of RFTS to two mono-ubiquitinated histone H3 disrupts the interaction between RFTS and the C-terminal catalytic domain, causing a conformational change in DNMT1 that allows hemi-methylated DNA to access the catalytic center. (iv) After the DNA substrate is fully methylated, DNMT1 is released from the ubiquitinated histone H3 to allow processive DNA methylation along the newly replicated DNA. Again, UHRF1 binds to hemi-methylated DNA first and adds ubiquitin marks to the neighboring histone H3. At the same time, the deubiquitinase USP7 is activated through a currently unknown mechanism to erase ubiquitin marks from histone H3 at methyl DNA fully recovered regions. Likely de-ubiquitination promotes the release of DNMT1 from histone H3 to ensure processive methylation reaction to proceed.

9. Future Perspective

The past several years have witnessed tremendous advances in our understanding of the regulatory mechanisms involved in maintenance DNA methylation. However, many questions remain to be addressed due to the complex interaction network of DNMT1, UHRF1 and other partner factors. UHRF1-dependent histone H3 ubiquitination is critical for maintenance DNA methylation, but it seems that ubiquitinated histone H3 exists transiently during replication, and USP7 has been suggested to be a histone H3 deubiquitinase [64]. It was previously demonstrated that USP7 promotes the stability of DNMT1 and UHRF1 through preventing two proteins from polyubiquitination and subsequent proteasomal degradation [65–67], while the role of USP7 in this process is still under debate. A recent study provides evidence suggesting USP7 is not required for DNMT1 stability and the interaction between USP7 and DNMT1 is unlikely to play a major role in DNMT1 homeostasis [68]. Nonetheless, a number of studies showed that USP7 physically interacts with DNMT1 and UHRF1 [51,66,69]. Binding of USP7 stimulates DNMT1 enzymatic activity in vitro [69]. Moreover, USP7 deubiquitinates the ubiquitinated histone H3 in vitro. Inhibition or depletion of USP7 causes accumulation of ubiquitinated histone H3 and compromises maintenance DNA methylation [64]. Given that the enzymatic activity of USP7 is allosterically regulated through the interaction with other partner proteins [70], it is important to determine how USP7 enzymatic activity is dynamically regulated during the process.

Recent evidence indicates that DNMT1 RFTS domain is a versatile protein interaction module. It has been demonstrated that DNMT1 RFTS domain interacts with the SRA domain and the interaction could stimulate DNMT1 enzymatic activity [29,35,69,71]. More recent studies found that the RFTS domain can interact with histone H3 N-terminal tail [27], but it binds much stronger to two mono-ubiquitinated histone H3 [57]. Given that maintenance DNA methylation is a highly efficient and orchestrated process, so far, more than 20 proteins have been reported to interact with DNMT1 [72]. Therefore, it is likely that some of these proteins, maybe in ubiquitinated form, could interact with RFTS domain to stimulate DNMT1 enzymatic activity during replication.

Lastly, Karg et al. have recently reported PAF15 is a major substrate for mono-ubiquitination by UHRF1 [73]. Dynamic PAF15 ubiquitination has previously been implicated in the regulation of translesion DNA bypass [74]. Interestingly, PAF15 exists in a mono-ubiquitinated form during normal replication, it will be interesting to explore if PAF15 has a role in maintenance DNA methylation.

Funding: The research in CM Qian's laboratory has been supported by Hong Kong Research Grants Council, General Research Fund 17160016, 17127715 and 17117514.

Acknowledgments: We thank Dr. Julian Tanner for critical reading of the manuscript.

Conflicts of Interest: The authors declare no conflict of interest.

References

1. Smith, Z.D.; Sindhu, C.; Meissner, A. Molecular features of cellular reprogramming and development. *Nat. Rev. Mol. Cell Biol.* **2016**, *17*, 139–154. [CrossRef]
2. Slotkin, R.K.; Martienssen, R. Transposable elements and the epigenetic regulation of the genome. *Nat. Rev. Genet.* **2007**, *8*, 272–285. [CrossRef] [PubMed]
3. Chang, S.C.; Tucker, T.; Thorogood, N.P.; Brown, C.J. Mechanisms of X-chromosome inactivation. *Front. Biosci.* **2006**, *11*, 852–866. [CrossRef] [PubMed]
4. Dodge, J.E.; Ramsahoye, B.H.; Wo, Z.G.; Okano, M.; Li, E. De novo methylation of MMLV provirus in embryonic stem cells: CpG versus non-CpG methylation. *Gene* **2002**, *289*, 41–48. [CrossRef]
5. Lister, R.; Pelizzola, M.; Dowen, R.H.; Hawkins, R.D.; Hon, G.; Tonti-Filippini, J.; Nery, J.R.; Lee, L.; Ye, Z.; Ngo, Q.M.; et al. Human DNA methylomes at base resolution show widespread epigenomic differences. *Nature* **2009**, *462*, 315–322. [CrossRef] [PubMed]
6. Deaton, A.M.; Bird, A. CpG islands and the regulation of transcription. *Genes Dev.* **2011**, *25*, 1010–1022. [CrossRef] [PubMed]
7. Okano, M.; Bell, D.W.; Haber, D.A.; Li, E. DNA methyltransferases Dnmt3a and Dnmt3b are essential for de novo methylation and mammalian development. *Cell* **1999**, *99*, 247–257. [CrossRef]
8. Kaneda, M.; Okano, M.; Hata, K.; Sado, T.; Tsujimoto, N.; Li, E.; Sasaki, H. Essential role for de novo DNA methyltransferase Dnmt3a in paternal and maternal imprinting. *Nature* **2004**, *429*, 900–903. [CrossRef]
9. Laurent, L.; Wong, E.; Li, G.; Huynh, T.; Tsirigos, A.; Ong, C.T.; Low, H.M.; Kin Sung, K.W.; Rigoutsos, I.; Loring, J.; et al. Dynamic changes in the human methylome during differentiation. *Genome Res.* **2010**, *20*, 320–331. [CrossRef]
10. Elliott, E.N.; Sheaffer, K.L.; Kaestner, K.H. The 'de novo' DNA methyltransferase Dnmt3b compensates the Dnmt1-deficient intestinal epithelium. *eLife* **2016**, *5*, e12975. [CrossRef]
11. Zemach, A.; McDaniel, I.E.; Silva, P.; Zilberman, D. Genome-wide evolutionary analysis of eukaryotic DNA methylation. *Science* **2010**, *328*, 916–919. [CrossRef] [PubMed]
12. Liang, G.; Chan, M.F.; Tomigahara, Y.; Tsai, Y.C.; Gonzales, F.A.; Li, E.; Laird, P.W.; Jones, P.A. Cooperativity between DNA methyltransferases in the maintenance methylation of repetitive elements. *Mol. Cell. Biol.* **2002**, *22*, 480–491. [CrossRef] [PubMed]
13. Ono, R.; Taki, T.; Taketani, T.; Taniwaki, M.; Kobayashi, H.; Hayashi, Y. LCX, leukemia-associated protein with a CXXC domain, is fused to MLL in acute myeloid leukemia with trilineage dysplasia having t(10;11)(q22;q23). *Cancer Res.* **2002**, *62*, 4075–4080. [PubMed]

14. Lorsbach, R.B.; Moore, J.; Mathew, S.; Raimondi, S.C.; Mukatira, S.T.; Downing, J.R. TET1, a member of a novel protein family, is fused to MLL in acute myeloid leukemia containing the t(10;11)(q22;q23). *Leukemia* **2003**, *17*, 637–641. [CrossRef] [PubMed]
15. Bostick, M.; Kim, J.K.; Esteve, P.O.; Clark, A.; Pradhan, S.; Jacobsen, S.E. UHRF1 plays a role in maintaining DNA methylation in mammalian cells. *Science* **2007**, *317*, 1760–1764. [CrossRef] [PubMed]
16. Sharif, J.; Muto, M.; Takebayashi, S.; Suetake, I.; Iwamatsu, A.; Endo, T.A.; Shinga, J.; Mizutani-Koseki, Y.; Toyoda, T.; Okamura, K.; et al. The SRA protein Np95 mediates epigenetic inheritance by recruiting Dnmt1 to methylated DNA. *Nature* **2007**, *450*, 908–912. [CrossRef] [PubMed]
17. Von Meyenn, F.; Iurlaro, M.; Habibi, E.; Liu, N.Q.; Salehzadeh-Yazdi, A.; Santos, F.; Petrini, E.; Milagre, I.; Yu, M.; Xie, Z.; et al. Impairment of DNA methylation maintenance is the main cause of global demethylation in naive embryonic stem cells. *Mol. Cell* **2016**, *62*, 848–861. [CrossRef]
18. Chuang, L.S.; Ian, H.I.; Koh, T.W.; Ng, H.H.; Xu, G.; Li, B.F. Human DNA-(cytosine-5) methyltransferase-PCNA complex as a target for p21WAF1. *Science* **1997**, *277*, 1996–2000. [CrossRef]
19. Rountree, M.R.; Bachman, K.E.; Baylin, S.B. DNMT1 binds HDAC2 and a new co-repressor, DMAP1, to form a complex at replication foci. *Nat. Genet.* **2000**, *25*, 269–277. [CrossRef]
20. Ding, F.; Chaillet, J.R. In vivo stabilization of the Dnmt1 (cytosine-5)-methyltransferase protein. *Proc. Natl. Acad. Sci. USA* **2002**, *99*, 14861–14866. [CrossRef]
21. Pradhan, M.; Esteve, P.O.; Chin, H.G.; Samaranayke, M.; Kim, G.D.; Pradhan, S. CXXC domain of human DNMT1 is essential for enzymatic activity. *Biochemistry* **2008**, *47*, 10000–10009. [CrossRef] [PubMed]
22. Song, J.; Rechkoblit, O.; Bestor, T.H.; Patel, D.J. Structure of DNMT1-DNA complex reveals a role for autoinhibition in maintenance DNA methylation. *Science* **2011**, *331*, 1036–1040. [CrossRef]
23. Zhang, Z.M.; Liu, S.; Lin, K.; Luo, Y.; Perry, J.J.; Wang, Y.; Song, J. Crystal structure of human DNA methyltransferase 1. *J. Mol. Biol.* **2015**, *427*, 2520–2531. [CrossRef] [PubMed]
24. Arita, K.; Isogai, S.; Oda, T.; Unoki, M.; Sugita, K.; Sekiyama, N.; Kuwata, K.; Hamamoto, R.; Tochio, H.; Sato, M.; et al. Recognition of modification status on a histone H3 tail by linked histone reader modules of the epigenetic regulator UHRF1. *Proc. Natl. Acad. Sci. USA* **2012**, *109*, 12950–12955. [CrossRef]
25. Avvakumov, G.V.; Walker, J.R.; Xue, S.; Li, Y.; Duan, S.; Bronner, C.; Arrowsmith, C.H.; Dhe-Paganon, S. Structural basis for recognition of hemi-methylated DNA by the SRA domain of human UHRF1. *Nature* **2008**, *455*, 822–825. [CrossRef] [PubMed]
26. Takeshita, K.; Suetake, I.; Yamashita, E.; Suga, M.; Narita, H.; Nakagawa, A.; Tajima, S. Structural insight into maintenance methylation by mouse DNA methyltransferase 1 (Dnmt1). *Proc. Natl. Acad. Sci. USA* **2011**, *108*, 9055–9059. [CrossRef]
27. Misaki, T.; Yamaguchi, L.; Sun, J.; Orii, M.; Nishiyama, A.; Nakanishi, M. The replication foci targeting sequence (RFTS) of DNMT1 functions as a potent histone H3 binding domain regulated by autoinhibition. *Biochem. Biophys. Res. Commun.* **2016**, *470*, 741–747. [CrossRef]
28. Bashtrykov, P.; Rajavelu, A.; Hackner, B.; Ragozin, S.; Carell, T.; Jeltsch, A. Targeted mutagenesis results in an activation of DNA methyltransferase 1 and confirms an autoinhibitory role of its RFTS domain. *ChemBioChem* **2014**, *15*, 743–748. [CrossRef] [PubMed]
29. Berkyurek, A.C.; Suetake, I.; Arita, K.; Takeshita, K.; Nakagawa, A.; Shirakawa, M.; Tajima, S. The DNA methyltransferase Dnmt1 directly interacts with the SET and RING finger-associated (SRA) domain of the multifunctional protein Uhrf1 to facilitate accession of the catalytic center to hemi-methylated DNA. *J. Biol. Chem.* **2014**, *289*, 379–386. [CrossRef]
30. Garvilles, R.G.; Hasegawa, T.; Kimura, H.; Sharif, J.; Muto, M.; Koseki, H.; Takahashi, S.; Suetake, I.; Tajima, S. Dual functions of the RFTS domain of Dnmt1 in replication-coupled DNA methylation and in protection of the genome from aberrant methylation. *PLoS ONE* **2015**, *10*, e0137509. [CrossRef] [PubMed]
31. Song, J.K.; Teplova, M.; Ishibe-Murakami, S.; Patel, D.J. Structure-based mechanistic insights into DNMT1-mediated maintenance DNA methylation. *Science* **2012**, *335*, 709–712. [CrossRef] [PubMed]
32. Hashimoto, H.; Horton, J.R.; Zhang, X.; Bostick, M.; Jacobsen, S.E.; Cheng, X. The SRA domain of UHRF1 flips 5-methylcytosine out of the DNA helix. *Nature* **2008**, *455*, 826–829. [CrossRef] [PubMed]
33. Arita, K.; Ariyoshi, M.; Tochio, H.; Nakamura, Y.; Shirakawa, M. Recognition of hemi-methylated DNA by the SRA protein UHRF1 by a base-flipping mechanism. *Nature* **2008**, *455*, 818–821. [CrossRef] [PubMed]
34. Qian, C.; Li, S.; Jakoncic, J.; Zeng, L.; Walsh, M.J.; Zhou, M.M. Structure and hemimethylated CpG binding of the SRA domain from human UHRF1. *J. Biol. Chem.* **2008**, *283*, 34490–34494. [CrossRef] [PubMed]

35. Bashtrykov, P.; Jankevicius, G.; Jurkowska, R.Z.; Ragozin, S.; Jeltsch, A. The UHRF1 protein stimulates the activity and specificity of the maintenance DNA methyltransferase DNMT1 by an allosteric mechanism. *J. Biol. Chem.* **2014**, *289*, 4106–4115. [CrossRef] [PubMed]
36. Nady, N.; Lemak, A.; Walker, J.R.; Avvakumov, G.V.; Kareta, M.S.; Achour, M.; Xue, S.; Duan, S.; Allali-Hassani, A.; Zuo, X.; et al. Recognition of multivalent histone states associated with heterochromatin by UHRF1 protein. *J. Biol. Chem.* **2011**, *286*, 24300–24311. [CrossRef]
37. Rajakumara, E.; Wang, Z.; Ma, H.; Hu, L.; Chen, H.; Lin, Y.; Guo, R.; Wu, F.; Li, H.; Lan, F.; et al. PHD finger recognition of unmodified histone H3R2 links UHRF1 to regulation of euchromatic gene expression. *Mol. Cell* **2011**, *43*, 275–284. [CrossRef]
38. Hu, L.; Li, Z.; Wang, P.; Lin, Y.; Xu, Y. Crystal structure of PHD domain of UHRF1 and insights into recognition of unmodified histone H3 arginine residue 2. *Cell Res.* **2011**, *21*, 1374–1378. [CrossRef]
39. Wang, C.; Shen, J.; Yang, Z.; Chen, P.; Zhao, B.; Hu, W.; Lan, W.; Tong, X.; Wu, H.; Li, G.; et al. Structural basis for site-specific reading of unmodified R2 of histone H3 tail by UHRF1 PHD finger. *Cell Res.* **2011**, *21*, 1379–1382. [CrossRef]
40. Xie, S.; Jakoncic, J.; Qian, C. UHRF1 double tudor domain and the adjacent PHD finger act together to recognize K9me3-containing histone H3 tail. *J. Mol. Biol.* **2012**, *415*, 318–328. [CrossRef]
41. Rothbart, S.B.; Krajewski, K.; Nady, N.; Tempel, W.; Xue, S.; Badeaux, A.I.; Barsyte-Lovejoy, D.; Martinez, J.Y.; Bedford, M.T.; Fuchs, S.M.; et al. Association of UHRF1 with methylated H3K9 directs the maintenance of DNA methylation. *Nat. Struct. Mol. Biol.* **2012**, *19*, 1155–1160. [CrossRef] [PubMed]
42. Liu, X.; Gao, Q.; Li, P.; Zhao, Q.; Zhang, J.; Li, J.; Koseki, H.; Wong, J. UHRF1 targets DNMT1 for DNA methylation through cooperative binding of hemi-methylated DNA and methylated H3K9. *Nat. Commun.* **2013**, *4*, 1563. [CrossRef] [PubMed]
43. Rothbart, S.B.; Dickson, B.M.; Ong, M.S.; Krajewski, K.; Houliston, S.; Kireev, D.B.; Arrowsmith, C.H.; Strahl, B.D. Multivalent histone engagement by the linked tandem Tudor and PHD domains of UHRF1 is required for the epigenetic inheritance of DNA methylation. *Genes Dev.* **2013**, *27*, 1288–1298. [CrossRef] [PubMed]
44. Citterio, E.; Papait, R.; Nicassio, F.; Vecchi, M.; Gomiero, P.; Mantovani, R.; Di Fiore, P.P.; Bonapace, I.M. Np95 is a histone-binding protein endowed with ubiquitin ligase activity. *Mol. Cell. Biol.* **2004**, *24*, 2526–2535. [CrossRef] [PubMed]
45. Karagianni, P.; Amazit, L.; Qin, J.; Wong, J. ICBP90, a novel methyl K9 H3 binding protein linking protein ubiquitination with heterochromatin formation. *Mol. Cell. Biol.* **2008**, *28*, 705–717. [CrossRef] [PubMed]
46. Nishiyama, A.; Yamaguchi, L.; Sharif, J.; Johmura, Y.; Kawamura, T.; Nakanishi, K.; Shimamura, S.; Arita, K.; Kodama, T.; Ishikawa, F.; et al. Uhrf1-dependent H3K23 ubiquitylation couples maintenance DNA methylation and replication. *Nature* **2013**, *502*, 249–253. [CrossRef] [PubMed]
47. Qin, W.; Wolf, P.; Liu, N.; Link, S.; Smets, M.; La Mastra, F.; Forne, I.; Pichler, G.; Horl, D.; Fellinger, K.; et al. DNA methylation requires a DNMT1 ubiquitin interacting motif (UIM) and histone ubiquitination. *Cell Res.* **2015**, *25*, 911–929. [CrossRef] [PubMed]
48. Gelato, K.A.; Tauber, M.; Ong, M.S.; Winter, S.; Hiragami-Hamada, K.; Sindlinger, J.; Lemak, A.; Bultsma, Y.; Houliston, S.; Schwarzer, D.; et al. Accessibility of different histone H3-binding domains of UHRF1 is allosterically regulated by phosphatidylinositol 5-phosphate. *Mol. Cell* **2014**, *54*, 905–919. [CrossRef]
49. Fang, J.; Cheng, J.; Wang, J.; Zhang, Q.; Liu, M.; Gong, R.; Wang, P.; Zhang, X.; Feng, Y.; Lan, W.; et al. Hemi-methylated DNA opens a closed conformation of UHRF1 to facilitate its histone recognition. *Nat. Commun.* **2016**, *7*, 11197. [CrossRef]
50. Harrison, J.S.; Cornett, E.M.; Goldfarb, D.; DaRosa, P.A.; Li, Z.M.; Yan, F.; Dickson, B.M.; Guo, A.H.; Cantu, D.V.; Kaustov, L.; et al. Hemi-methylated DNA regulates DNA methylation inheritance through allosteric activation of H3 ubiquitylation by UHRF1. *eLife* **2016**, *5*, e17101. [CrossRef]
51. Zhang, Z.M.; Rothbart, S.B.; Allison, D.F.; Cai, Q.; Harrison, J.S.; Li, L.; Wang, Y.; Strahl, B.D.; Wang, G.G.; Song, J. An allosteric interaction links USP7 to deubiquitination and chromatin targeting of UHRF1. *Cell Rep.* **2015**, *12*, 1400–1406. [CrossRef] [PubMed]
52. Gao, L.; Tan, X.F.; Zhang, S.; Wu, T.; Zhang, Z.M.; Ai, H.W.; Song, J. An intramolecular interaction of UHRF1 reveals dual control for Its histone association. *Structure* **2018**, *26*, 304–311. [CrossRef] [PubMed]
53. Ferry, L.; Fournier, A.; Tsusaka, T.; Adelmant, G.; Shimazu, T.; Matano, S.; Kirsh, O.; Amouroux, R.; Dohmae, N.; Suzuki, T.; et al. Methylation of DNA ligase 1 by G9a/GLP recruits UHRF1 to replicating DNA and regulates DNA methylation. *Mol. Cell* **2017**, *67*, 550–565. [CrossRef] [PubMed]

54. Vaughan, R.M.; Dickson, B.M.; Whelihan, M.F.; Johnstone, A.L.; Cornett, E.M.; Cheek, M.A.; Ausherman, C.A.; Cowles, M.W.; Sun, Z.W.; Rothbart, S.B. Chromatin structure and its chemical modifications regulate the ubiquitin ligase substrate selectivity of UHRF1. *Proc. Natl. Acad. Sci. USA* **2018**, *115*, 8775–8780. [CrossRef] [PubMed]
55. DaRosa, P.A.; Harrison, J.S.; Zelter, A.; Davis, T.N.; Brzovic, P.; Kuhlman, B.; Klevit, R.E. A Bifunctional role for the UHRF1 UBL domain in the control of hemi-methylated DNA-dependent histone ubiquitylation. *Mol. Cell* **2018**, *72*, 753–765. [CrossRef] [PubMed]
56. Foster, B.M.; Stolz, P.; Mulholland, C.B.; Montoya, A.; Kramer, H.; Bultmann, S.; Bartke, T. Critical role of the UBL domain in stimulating the E3 ubiquitin ligase activity of UHRF1 toward chromatin. *Mol. Cell* **2018**, *72*, 739–752. [CrossRef] [PubMed]
57. Ishiyama, S.; Nishiyama, A.; Saeki, Y.; Moritsugu, K.; Morimoto, D.; Yamaguchi, L.; Arai, N.; Matsumura, R.; Kawakami, T.; Mishima, Y.; et al. Structure of the Dnmt1 reader module complexed with a unique two-mono-ubiquitin mark on histone H3 reveals the basis for DNA methylation maintenance. *Mol. Cell* **2017**, *68*, 350–360. [CrossRef]
58. Li, T.; Wang, L.; Du, Y.; Xie, S.; Yang, X.; Lian, F.; Zhou, Z.; Qian, C. Structural and mechanistic insights into UHRF1-mediated DNMT1 activation in the maintenance DNA methylation. *Nucleic Acids Res.* **2018**, *46*, 3218–3231. [CrossRef]
59. Zhao, Q.; Zhang, J.; Chen, R.; Wang, L.; Li, B.; Cheng, H.; Duan, X.; Zhu, H.; Wei, W.; Li, J.; et al. Dissecting the precise role of H3K9 methylation in crosstalk with DNA maintenance methylation in mammals. *Nat. Commun.* **2016**, *7*, 12464. [CrossRef]
60. Cheng, J.; Yang, Y.; Fang, J.; Xiao, J.; Zhu, T.; Chen, F.; Wang, P.; Li, Z.; Yang, H.; Xu, Y. Structural insight into coordinated recognition of trimethylated histone H3 lysine 9 (H3K9me3) by the plant homeodomain (PHD) and tandem tudor domain (TTD) of UHRF1 (ubiquitin-like, containing PHD and RING finger domains, 1) protein. *J. Biol. Chem.* **2013**, *288*, 1329–1339. [CrossRef]
61. Veland, N.; Hardikar, S.; Zhong, Y.; Gayatri, S.; Dan, J.; Strahl, B.D.; Rothbart, S.B.; Bedford, M.T.; Chen, T. The arginine methyltransferase PRMT6 regulates DNA methylation and contributes to global DNA hypomethylation in cancer. *Cell Rep.* **2017**, *21*, 3390–3397. [CrossRef] [PubMed]
62. Sirbu, B.M.; Couch, F.B.; Feigerle, J.T.; Bhaskara, S.; Hiebert, S.W.; Cortez, D. Analysis of protein dynamics at active, stalled, and collapsed replication forks. *Genes Dev.* **2011**, *25*, 1320–1327. [CrossRef] [PubMed]
63. Robertson, K.D.; Ait-Si-Ali, S.; Yokochi, T.; Wade, P.A.; Jones, P.L.; Wolffe, A.P. DNMT1 forms a complex with Rb, E2F1 and HDAC1 and represses transcription from E2F-responsive promoters. *Nat. Genet.* **2000**, *25*, 338–342. [CrossRef] [PubMed]
64. Yamaguchi, L.; Nishiyama, A.; Misaki, T.; Johmura, Y.; Ueda, J.; Arita, K.; Nagao, K.; Obuse, C.; Nakanishi, M. Usp7-dependent histone H3 deubiquitylation regulates maintenance of DNA methylation. *Sci. Rep.* **2017**, *7*, 55. [CrossRef] [PubMed]
65. Du, Z.; Song, J.; Wang, Y.; Zhao, Y.; Guda, K.; Yang, S.; Kao, H.Y.; Xu, Y.; Willis, J.; Markowitz, S.D.; et al. DNMT1 stability is regulated by proteins coordinating deubiquitination and acetylation-driven ubiquitination. *Sci. Signal.* **2010**, *3*, ra80. [CrossRef] [PubMed]
66. Cheng, J.; Yang, H.; Fang, J.; Ma, L.; Gong, R.; Wang, P.; Li, Z.; Xu, Y. Molecular mechanism for USP7-mediated DNMT1 stabilization by acetylation. *Nat. Commun.* **2015**, *6*, 7023. [CrossRef] [PubMed]
67. Ma, H.; Chen, H.; Guo, X.; Wang, Z.; Sowa, M.E.; Zheng, L.; Hu, S.; Zeng, P.; Guo, R.; Diao, J.; et al. M phase phosphorylation of the epigenetic regulator UHRF1 regulates its physical association with the deubiquitylase USP7 and stability. *Proc. Natl. Acad. Sci. USA* **2012**, *109*, 4828–4833. [CrossRef]
68. Yarychkivska, O.; Tavana, O.; Gu, W.; Bestor, T.H. Independent functions of DNMT1 and USP7 at replication foci. *Epigenet. Chromatin* **2018**, *11*, 9. [CrossRef]
69. Felle, M.; Joppien, S.; Nemeth, A.; Diermeier, S.; Thalhammer, V.; Dobner, T.; Kremmer, E.; Kappler, R.; Langst, G. The USP7/Dnmt1 complex stimulates the DNA methylation activity of Dnmt1 and regulates the stability of UHRF1. *Nucleic Acids Res.* **2011**, *39*, 8355–8365. [CrossRef]
70. Kim, R.Q.; Sixma, T.K. Regulation of USP7: A high incidence of E3 complexes. *J. Mol. Biol.* **2017**, *429*, 3395–3408. [CrossRef]
71. Achour, M.; Jacq, X.; Ronde, P.; Alhosin, M.; Charlot, C.; Chataigneau, T.; Jeanblanc, M.; Macaluso, M.; Giordano, A.; Hughes, A.D.; et al. The interaction of the SRA domain of ICBP90 with a novel domain of DNMT1 is involved in the regulation of VEGF gene expression. *Oncogene* **2008**, *27*, 2187–2197. [CrossRef] [PubMed]

72. Qin, W.; Leonhardt, H.; Pichler, G. Regulation of DNA methyltransferase 1 by interactions and modifications. *Nucleus* **2011**, *2*, 392–402. [CrossRef]
73. Karg, E.; Smets, M.; Ryan, J.; Forne, I.; Qin, W.; Mulholland, C.B.; Kalideris, G.; Imhof, A.; Bultmann, S.; Leonhardt, H. Ubiquitome analysis reveals PCNA-associated factor 15 (PAF15) as a specific ubiquitination target of UHRF1 in embryonic stem cells. *J. Mol. Biol.* **2017**, *429*, 3814–3824. [CrossRef] [PubMed]
74. Povlsen, L.K.; Beli, P.; Wagner, S.A.; Poulsen, S.L.; Sylvestersen, K.B.; Poulsen, J.W.; Nielsen, M.L.; Bekker-Jensen, S.; Mailand, N.; Choudhary, C. Systems-wide analysis of ubiquitylation dynamics reveals a key role for PAF15 ubiquitylation in DNA-damage bypass. *Nat. Cell Biol.* **2012**, *14*, 1089–1098. [CrossRef] [PubMed]

© 2018 by the authors. Licensee MDPI, Basel, Switzerland. This article is an open access article distributed under the terms and conditions of the Creative Commons Attribution (CC BY) license (http://creativecommons.org/licenses/by/4.0/).

Review

Coordinated Dialogue between UHRF1 and DNMT1 to Ensure Faithful Inheritance of Methylated DNA Patterns

Christian Bronner [1,*], Mahmoud Alhosin [2], Ali Hamiche [1] and Marc Mousli [3]

[1] Institut de Génétique et de Biologie Moléculaire et Cellulaire (IGBMC), Université de Strasbourg, INSERM U1258 CNRS UMR 7104, 1 rue Laurent Fries, 67404, Illkirch, France; hamiche@igbmc.fr
[2] Department of Biochemistry, Faculty of Science and Cancer and Mutagenesis Unit, King Fahd Medical Research Center, King Abdulaziz University, Jeddah, Saudi Arabia; malhaseen@kau.edu.sa
[3] Laboratoire de Bioimagerie et Pathologies, Faculté de Pharmacie, Université de Strasbourg, CNRS UMR 7021, 67401 Illkirch, France; marc.mousli@unistra.fr
* Correspondence: bronnerc@igbmc.fr; Tel.: +33-88-65-33-48

Received: 26 November 2018; Accepted: 11 January 2019; Published: 18 January 2019

Abstract: DNA methylation, catalyzed by DNA methyltransferases (DNMTs), is an epigenetic mark that needs to be faithfully replicated during mitosis in order to maintain cell phenotype during successive cell divisions. This epigenetic mark is located on the 5′-carbon of the cytosine mainly within cytosine–phosphate–guanine (CpG) dinucleotides. DNA methylation is asymmetrically positioned on both DNA strands, temporarily generating a hemi-methylated state after DNA replication. Hemi-methylation is a particular status of DNA that is recognized by ubiquitin-like containing plant homeodomain (PHD) and really interesting new gene (RING) finger domains 1 (UHRF1) through its SET- (**S**u(var)3-9, **E**nhancer-of-zeste and **T**rithorax) and RING-associated (SRA) domain. This interaction is considered to be involved in the recruitment of DNMT1 to chromatin in order to methylate the adequate cytosine on the newly synthetized DNA strand. The UHRF1/DNMT1 tandem plays a pivotal role in the inheritance of DNA methylation patterns, but the fine-tuning mechanism remains a mystery. Indeed, because DNMT1 experiences difficulties in finding the cytosine to be methylated, it requires the help of a guide, i.e., of UHRF1, which exhibits higher affinity for hemi-methylated DNA vs. non-methylated DNA. Two models of the UHRF1/DNMT1 dialogue were suggested to explain how DNMT1 is recruited to chromatin: (i) an indirect communication via histone H3 ubiquitination, and (ii) a direct interaction of UHRF1 with DNMT1. In the present review, these two models are discussed, and we try to show that they are compatible with each other.

Keywords: cell identity; DNA methylation; DNMT1; epigenetics; gene expression; UHRF1

---

1. Introduction

1.1. DNA Methylation Patterns: Layers of Epigenomes

DNA methylation mainly occurs on cytosine–phosphate–guanine (CpG) dinucleotides, and almost 70–80% of all CpG dinucleotides (~3 × 10^7) are methylated in human cell genomes [1,2]. DNA methylation can also occur at non-CpG, including cytosine–phosphate–adenine (CpA), cytosine–phosphate–thymine (CpT), and cytosine–phosphate–cytosine (CpC), with the first one being the most frequent [3]. The role of DNA methylation is still a matter of debate [3]. Considering that de novo DNA methyltransferases (DNMTs), such as DNMT3A and DNMT3B, catalyze methylation at non-CpG sites, we do not extend this review to non-CpG methylation. Although the role of DNA methylation was reviewed elsewhere, before going deeper into the insights of the mechanism of DNA methylation pattern duplication, a few reminders on the location of the methylated CpG sites are made hereafter.

Methylated CpG dinucleotides are preferentially localized within repetitive elements (SINEs; Short Interspersed Nuclear Elements, LINEs; Long INterspersed Nuclear Elements, and LTR; Long Terminal Repeat), centromeres, and coding regions of functional genes (gene bodies) [4]. Methylation of CpG dinucleotides in these areas guarantees genomic stability and prevents aberrant initiation of transcription [5,6]. The primary well-recognized role of DNA methylation lies in transcriptional repression of retrotransposons, X chromosome inactivation, and mono-allelic expression of imprinted genes [7]. Therefore, the methylation of DNA, along with enhancers and gene bodies [8–10], is involved in gene expression regulation. Non-methylated CpG dinucleotides are frequently observed in promoter regions of active genes where they form "CpG islands (CGIs)". CpG islands are regions greater than 550 bp of DNA with a GC content greater than 50%, and an observed CpG/expected CpG ratio greater than 0.6 [11]. Approximately, 70% of human gene promoters contain CGIs, usually unmethylated in normal cells, thus allowing transcription of the corresponding gene. During development and differentiation, around 6% of them undergo methylation in a tissue-specific manner, thereby contributing to the occurrence of different phenotypes [5,11]. Irregularities in methylation pattern, especially in promoters, can lead to profound abnormalities including malignant transformation of cells [4]. In cancers, two opposite patterns of DNA methylation are observed and contribute together to tumorigenesis. Firstly, CpG islands in promoter regions of various tumor suppressor genes (TSGs), such as $p16^{INK4A}$, *p73*, *BRCA1*, and *TIMP3*, are hypermethylated in cancer, inhibiting their expression and, thus, leading to uncontrolled proliferation of cancer cells [12–15]. Nonetheless, establishment of this methylation pattern is not yet deciphered. Secondly, a global hypomethylation is observed in cancer cells, which promotes the expression of oncogenes and aggravates tumorigenesis by inducing genomic instability [16,17]. The origin of these two upheavals in DNA methylation patterns is not yet fully elucidated, but might find a beginning of an explanation in the altered communication between ubiquitin-like containing plant homeodomain (PHD) and really interesting new gene (*RING*) finger domains 1 (UHRF1) and DNA methyltransferase 1 (DNMT1) [18,19], which is the matter of the present review. However, what is clear is that these DNA methylation patterns are faithfully transmitted from a mother cancer cell to the two daughter cancer cells with a phenotype preserved almost to infinity. Considering that cancer cells usually divide faster than normal cells and that they have a more stable phenotype throughout successive mitosis, especially under in vitro culture conditions [20], one might imagine that the DNMT1/UHRF1 tandem is more efficient in cancer cells than non-cancerous cells.

Efforts over the past decade shed light onto the dynamics and the molecular mechanisms underlying DNA methylation pattern duplication occurring after DNA replication. One important breakthrough was achieved by the discovery of the UHRF1/DNMT1 network.

1.2. The Role of UHRF1/DNMT1 Tandem

Chromosomal mapping of human *UHRF1* and *DNMT1* genes both at 19p13, more specifically at 19p13.3 and 19p13.2, respectively, [21] and genome sequencing revealed that they are separated by about 50 Mb, i.e., 50 centimorgans (personal observations). It is worth noting that the tissue-specific expression of UHRF1 and DNMT1 is tightly linked, suggesting that they need each other to exert their role. Indeed, for both genes, the most elevated expressions were found in the appendix, bone marrow, lymph node, and testis [22,23]. In contrast, highly differentiated tissues, such as the heart, liver, pancreas, prostate, and salivary glands, were among the tissues that express the lowest levels of *DNMT1* and *UHRF1* messenger RNAs (mRNAs) [22,23]. This may explain that UHRF1$^{-/-}$ phenocopies DNMT1$^{-/-}$ [24,25], that DNMT$^{-/-}$ is embryonic lethal at day 9 [26], and that UHRF1$^{-/-}$ is mid-gestational lethal [27], suggesting that both genes have interdependent roles in DNA methylation maintenance. This also suggests that both genes might have similar pathways of regulation. Pharmacological or pathological regulation of the expression of *UHRF1* and *DNMT1* genes was extensively reviewed elsewhere [22,28–37]. Briefly, one interesting point with deep impact is that downregulation of *UHRF1* and/or *DNMT1* always allows re-expression or enhanced expression of a large number of tumor suppressor genes, including *RB1*, *p16^{INK4A}*, *CDH13*, *SHP1*, *SOCS3*, *3OST2*, *BRCA1*, *CDX2*, *RUNX3*, *FOXO4*, *PPARG*, *PML*, *MEG3*, *HHIP*, *IGFBP3*, *SFRP1*, and *14-3-3σ* [29,34,35,38–45]. Also of note, it was shown that UHRF1 is involved in epigenetic silencing of *KiSS1*, a metastasis suppressor gene [46]. The mechanism of the re-expression of tumor suppressor genes following decreased expression of *DNMT1* and *UHRF1* remains elusive. Indeed, it is clear that these two proteins are involved in the maintenance of hypermethylation of promoters, but how they are demethylated remains a mystery. A passive demethylation, via a downregulation of *UHRF1*, through successive mitosis appears unlikely, as *UHRF1* is indispensable for cell proliferation [22,47].

For global DNA hypomethylation, a diminished interaction was suggested between *UHRF1* and *DNMT1* [18,19], although *UHRF1* expression is enhanced in all cancers so far investigated [29,34–36,48]. However, a ubiquitin-dependent degradation of *DNMT3A* induced by *UHRF1* and/or *UHRF2* might also be involved [49]. Indeed, considering that DNMT3A is involved in de novo DNA methylation, an increase of UHRF1 through the targeting of DNMT3A also likely contributes to the global DNA hypomethylation in cancer cells [49].

The maintenance of DNA methylation at the replication fork is believed to be ensured by the DNMT1/PCNA (Proliferating Cell Nuclear Antigen) tandem [50–53]; however, surprisingly, its disruption exerts little effect on genomic DNA methylation in contrast to that of the UHRF1/DNMT1 tandem, which induces massive DNA hypomethylation [18]. Furthermore, the occurrence of a disruption in the oncogenic process was extended to the UHRF1/DNMT1/PCNA complex [19]. Complementary to this event, it was demonstrated that UHRF1 overexpression drives DNA hypomethylation by delocalizing DNMT1 [54], which further supports that abnormal cooperation within the UHRF1/DNMT1 tandem may be one of the first steps of tumorigenesis onset. Indeed, the de novo methylation of genes frequently observed in cancers could be catalyzed by DNMT1, rather than by DNMT3A or DNMT3B [26,55,56].

At the molecular level, the role of UHRF1 can be summarized in the targeting of DNMT1 to replication forks by serving as a guide for DNMT1 at hemi-methylated CpG sites [25,57–64]. At the cellular level, the UHRF1/DNMT1 tandem is involved in many processes, including differentiation [65], cell senescence [66], stem cell self-renewal [67–69], neurogenesis [70], germinal center B-cell expansion [71], maturation of colonic T lymphocytes [72], smooth muscle plasticity [73], and induced pluripotent stem-cell reprogramming [74] and development [75,76].

1.3. UHRF1 and DNMT1, Interdependent Multi-Domain Proteins

UHRF1 is a multi-domain protein (Figure 1) including a ubiquitin-like domain (UBL), a tandem Tudor domain (TTD), a plant homeodomain (PHD), an SET- and RING-associated (SRA) domain, and a really interesting new gene (RING) domain, with the latter domain conferring the sole enzymatic activity [77]. Two-thirds of the primary sequence of UHRF1 contributes to these structural domains, suggesting that this protein fulfils multiple important roles. One of these roles is the link between the histone code and DNA methylation. Indeed, SRA is able to sense the presence of hemi-methylated DNA through its SRA domain and to recognize via the tandem Tudor domain, probably at the same time, two or three methyl groups on lysine 9 of histone H3 (H3K9me2/3) [25,61,64,78–82]. Accordingly, it was shown that UHRF1 can target DNMT1 for DNA methylation maintenance via binding to H3K9me2/3 or hemi-methylated CpG [83], thus ensuring a kind of security for the faithful inheritance of DNA methylation and, probably reciprocally, histone H3 methylation profiles. The PHD domain is involved in binding to unmodified H3R2 [84–87] and, thus, contributes to the binding of UHRF1 to chromatin [88]. The role of the UBL might allow UHRF1 to follow the proteasome pathway [89] and/or be involved in the interaction with the E2 conjugate to promote E3 ligase activity [90,91].

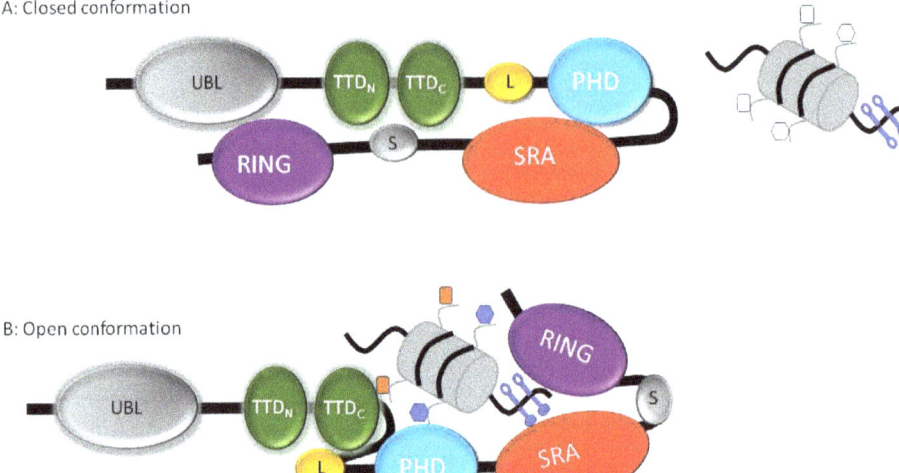

Figure 1. Schematic representation of ubiquitin-like containing plant homeodomain (PHD) and really interesting new gene (RING) finger domains 1 (UHRF1)'s domains with open and closed conformations. (**a**) Closed conformation of UHRF1. In the absence of methylated DNA (open lollipops), UHRF1 adopts a closed conformation with the spacer (S) facing the tandem Tudor domain (TTD) and the RING domain facing the ubiquitin-like domain (UBL). This interaction may represent a kind of auto-inhibitory mechanism as UBL exhibits analogy with ubiquitin. The SET- and RING-associated (SRA) domain faces the PHD domain. The linker (L) is the sequence between the TTD and PHD. (**b**) Open conformation of UHRF1. In the presence of hemi-methylated DNA (full lollipops on one strand) and in the presence of di/trimethylated lysine 9 on histone H3 (ocher rectangle on nucleosome) and of unmodified arginine 2 on histone H3 (blue hexagon on nucleosome), UHRF1 adopts an open conformation allowing each domain to fulfill its respective role. For instance, the RING domain can ubiquitinate histone H3 and the TTD with methylated histone H3.

DNMT1 is a large protein of 1616 amino acids (aa) [31,92–95] composed of a large regulatory N-terminal region (1000 aa) and a catalytic C-terminal region including 10 catalytic motifs that are important for the interaction with S-adenosyl methionine, the methyl group donor. The C-terminal and N-terminal regions are linked by six Lys/Gly (KG) dipeptide repeats. The N-terminal region is composed of (i) a domain able to interact with a large panel of proteins, (ii) a replication focus targeting sequence (RFTS) domain, which is a matter of very active research over the last few years, involved in the recruitment of DNMT1 to the DNA replication fork, (iii) a zinc-binding domain, (iv) two bromo-adjacent homology (BAH) domains, and (v) a nuclear localization signal (aa 191–211). The crystal structure revealed that the RFTS domain directly associates with the catalytic domain, thereby exhibiting auto-inhibition ability [96].

2. The UHRF1/DNMT1 Dialogue on Chromatin

2.1. The Ubiquitination of Histone H3 by UHRF1 as a Chromatin Anchorage for DNMT1: Model A

It was previously shown by several groups [96–99] that UHRF1 adopts a closed conformation with a C-terminal part (spacer aa 586–674) facing the TTD, and the SRA domain facing the PHD domain (Figure 1a). This spacer region contains a polybasic region (PBR) in the C-terminus, which prevents the recognition of H3K9me3 by the TTD domain. Therefore, in this conformation, it appears that UHRF1 is unable to bind to H3K9me2/3 via its TTD, to the unmodified R2 of histone H3 via the PHD domain, or to hemi-methylated DNA via its SRA domain [97]. However, interestingly, the presence of hemi-methylated DNA impairs the intramolecular interactions, and each domain can exert its ability to interact with its selective target (Figure 1b). It was further suggested that the docking of TTD to H3K9me3 may induce the ubiquitination activity of the RING domain toward lysines 14, 18, and 23 (K14, K18, K23) of histone H3 [97], thus giving rise to a signal for DNMT1 recruitment [100–103]. Fine-tuned regulation of H3 ubiquitination suggests that the RING finger of UHRF1 induces mono-ubiquitination of histone H3 at K14, K18, and K23 but not polyubiquitination [88,102] in spite of its ability to catalyze it [104,105]. Apparently, two mono-ubiquitinated lysines are enough for DNMT1 to bind histone H3, i.e., H3K14/K18, H3K14/K18, or H3K18/K23, with the help of the N-terminal tail of histone H3 to confer high affinity [102]. It is the RFTS domain of DNMT1 that is involved in this anchorage with similar affinities for each double mono-ubiquitinated motif [102]. The mode of ubiquitin binding is, however, different from other reported recognition patterns [106,107]. Regarding the high affinity of DNMT1 for ubiquitinated histone H3, it was suggested that this interaction gives robustness to the inheritance of DNA methylation patterns [106]. It was further proposed that these lysines could be crucial for gene expression regulation as they can be ubiquitinated (repressive effect) or acetylated by p300/CBP or GCN5 (General Control Of Amino Acid Synthesis Protein 5) which are enzymes involved in gene transcription [108].

In accordance with a role of histone H3 ubiquination in DNA methylation maintenance, depletion of ubiquitin-specific peptidase 7 (USP7) in HeLa cells enhanced histone H3 ubiquitination and led to enhancement of DNMT1 nuclear foci during DNA replication [109]. However, it was shown that USP7 stimulates both the maintenance and de novo DNA methylation activity of DNMT1 in vitro [88,110], further supporting a complex role of USP7 in DNA methylation maintenance that remains to be fully deciphered.

2.2. Domain–Domain Interactions between DNMT1 and UHRF1: Model B

There were numerous reports [24,25,60,111–113] showing a physical interaction between UHRF1 and DNMT1, leaving little doubt that UHRF1 communicates directly with DNMT1. Initially, two independent groups reported the interaction between UHRF1 and DNMT1 [24,60]. Domain mapping nevertheless showed some discrepancies. Indeed, we observed that the human SRA domain of UHRF1 is responsible for the interaction with human DNMT1 [60] (Figure 2b). The other group found that it is rather the PHD domain of mouse UHRF1 that is involved in DNMT1 interaction [24]. Additionally, it was recently reported that the UBL domain of UHRF1 is able to bind to the RFTS domain of DNMT1 (N-terminal region), as well with the amino-acid sequence encompassing aa 621–1616 [91,106]. The latter interaction is involved in the stimulation of the enzymatic activity of DNMT1, whereas the first (RFTS) is rather implicated in the alleviation of the auto-inhibition exerted by DNMT1 [91,106]. Three regions of DNMT1 were shown to bind to UHRF1, namely amino-acid residues 1–446, 1081–1408 in mouse [24], and 401–615 in human [60]. We previously suggested that the discrepancies might come from the species-specific difference [59]; however, it was later shown that the interacting domain in mouse DNMT1 was a region encompassing aa 291–601 corresponding to the RFTS domain [114]. In accordance, it was shown that the interacting motif of DNMT1 with UHRF1 is a region spanning from aa 380–399 [88]. Interestingly and consistently, the spacer of UHRF1, located beside the SRA domain, finds an important role as it facilitates the interaction of the SRA domain with hemi-methylated DNA and the interaction with DNMT1 [97]. It should be mentioned here that the spacer might explain the different functions of UHRF1 and UHRF2, as the latter lacks this sequence, as well as DNA methylation maintenance capacity [111,115,116]. Therefore, the most likely sites of interaction are the SRA domain and the UBL of UHRF1, with the RFTS domain of DNMT1 (Figure 2). This multiple site connection might establish a kind of dialogue between these two proteins depending on the methylation status of the DNA and/or the motion of the complex along the DNA. In accordance with this, it was shown that full-length UHRF1, the SRA domain [117], or the UBL [106] increase the preference or catalytic activity of DNMT1 for hemi-methylated DNA. This effect is dependent on the RFTS domain of DNMT1. Indeed, the RFTS of DNMT1 exerts inhibitory action on its catalytic domain, by occupying the DNA binding pocket, as shown by crystallography [112,113]. This could be evidenced by incubating DNMT1 with UHRF1 or the SRA domain prior to the addition of DNA, as UHRF1 competes with DNMT1 for binding to DNA. When DNA is added first, UHRF1 binds to DNA and does not reverse the clamped conformation of DNMT1. A model in which UHRF1 binds to hemi-methylated CpG site via its SRA domain with subsequent recruitment of DNMT1 was suggested [117]. This model supposes that UHRF1, through the SRA domain, binds firstly to DNA, which is in accordance with other studies proposing that the SRA domain moves along DNA to seek for the presence of hemi-methylated DNA [57,58]. In accordance with a histone-independent recruitment of DNMT1, it was intriguingly proposed that auto-ubiquitination of UHRF1 might serve as a docking site for DNMT1 [118].

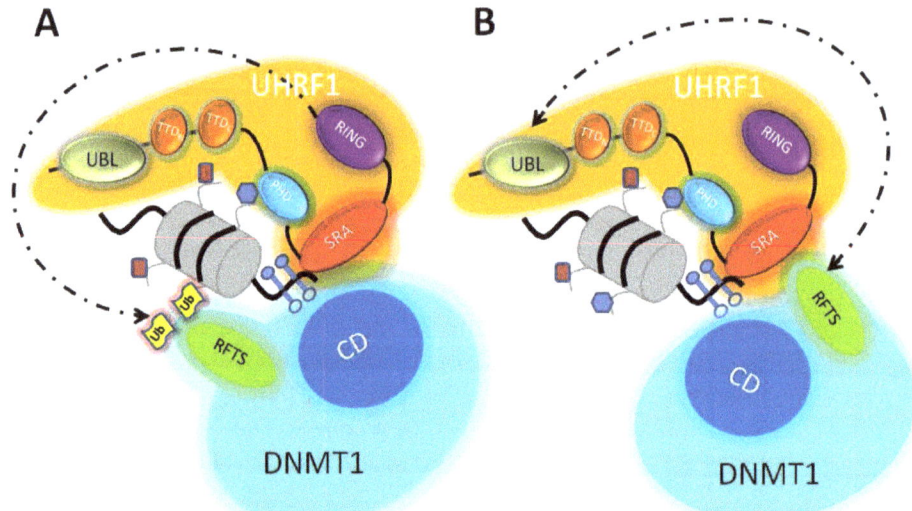

Figure 2. Dialogue model between UHRF1 and DNA methyltransferase 1 (DNMT1) in the presence of hemi-methylated DNA. (**a**) Model A: targeting of DNMT1 to chromatin via histone H3-dependent ubiquitination. Two molecules of ubiquitin (Ub) on the nucleosome, mediated by the RING domain of UHRF1 (dotted line), serve as an anchorage for DNMT1 via the replication focus targeting sequence (RFTS) domain. This interaction allows the alleviation of the auto-inhibitory activity of RFTS on the catalytic domain (CD) of DNMT1. Concomitantly, the TTD binds di/trimethylated lysine 9 on histone H3 (ocher rectangle on nucleosome) and the PDH binds unmodified arginine 2 on histone H3 (blue hexagon on nucleosome). (**b**) Model B: targeting of DNMT1 to chromatin via UHRF1 domain interactions. The UBL (dotted line) and SRA domain interact with the RFTS domain of UHRF1, allowing the release of the catalytic domain (CD) of DNMT1. As with model A, the TTD binds di/trimethylated lysine 9 on histone H3 (ocher rectangle on nucleosome) and the PDH binds unmodified arginine 2 on histone H3 (blue hexagon on nucleosome).

2.3. A Conciliated Model of How the UHRF1/DNMT1 Tandem Works

In light of all the studies, the following question remains: what is the correct model for the UHRF1-dependent CpG targeting by DNMT1? The majority of these studies were performed on purified isolated proteins, thus questioning the relevance of the *in vivo* occurring. However, if both models are right, we have to find an explanation to render them compatible.

We, therefore, propose a more complete model in which UHRF1 does not recruit DNMT1, but in which they are still together in the same large macro-molecular complex; thus, DNMT1 is in permanent presence, in close vicinity of UHRF1, as long as this complex progresses at the DNA replication fork, although a direct transfer of a CpG site from the SRA to DNMT1 appears not applicable [117]. When the SRA domain meets a hemi-methylated CpG dinucleotide, UHRF1 adopts an open conformation allowing ubiquitination of histone H3 and/or auto-ubiquitination via the RING, facilitated by the cooperation of the UBL and E2 Ube2D (ubiquitin-conjugating enzyme E2D). The ubiquitination of histone H3 and/or UHRF1 auto-ubiquitination represent an anchorage signal for the RFTS domain of DNMT1, liberating the catalytic activity of DNMT1. Meanwhile the UHRF1/TTD tandem binds to the methylated histone H3 at lysine 9. This first step is preparing DNMT1 to enter in action, i.e., to methylate the new DNA strand; however, this is not sufficient to localize it. We further propose that UHRF1, when flipping the mC through the SRA domain may undergo an allosteric change that allows creating contact zones with DNMT1 inside the large complex they belong to. This interaction facilitates accession of the catalytic center of DNMT1 to its substrate, i.e., to hemi-methylated DNA, along with the release to mC by the SRA domain, considering that steric hindrance forbids that both mCs on each DNA strand can be flipped at the same time by UHRF1 and DNMT1 [119]. We do not exclude that, when UHRF1 releases the mC, it adopts a conformational change which promotes the seeking of other hemi-methylated CpG sites, leaving behind DNMT1 to methylate the newly synthesized DNA strand. We do not exclude either that the RFTS domain, chronologically, may firstly interact with ubiquitinated histone H3 and, secondly, with the SRA domain in order to remove UHRF1 from the CpG dinucleotide, considering that DNMT1 cannot methylate the opposite DNA strand as long as UHRF1 is present on the same targeted hemi-methylated CpG.

Our model supports a repetitive dialogue between DNMT1 and UHRF1 during the synthesis (S) phase at CpG sites, and is in opposition to a model in which DNMT1 leaves the chromatin as soon as it catalyzes one enzymatic reaction, i.e., a hydrogen atom replaced by a methyl group on a cytosine. This model is particularly likely when CpG islands (repeated methylated CpG) are methylated. An interesting question arises from this model: what happens between two methylated CpG dinucleotides or between two methylated CpG islands? In our opinion, it is unlikely that DNMT1 dissociates from UHRF1 or the UHRF1 complex after each hemi-methylated CpG dinucleotide. Otherwise, it would take too much time for UHRF1 to wait for recruiting (i.e., to find) a molecule of DNMT1; furthermore, what would be the signal telling DNMT1 to join the UHRF1 complex? All that would hinder the epigenetic code replication machinery, in which UHRF1 plays the role of conductor, to follow the progression of the DNA replication fork as fast as required. In contrast, a more plausible behavior of DNMT1 and UHRF1 would be that they are in close contact in the complex while moving along the DNA, and when UHRF1, through the SRA domain, flips out the methylated cytosine, it probably communicates with DNMT1 to prepare it for recognition of the unmethylated cytosine on the opposite DNA strand. Due to allosteric constraints, UHRF1 needs to further progress to allow DNMT1 to exert its catalytic activity. Accordingly, it was proposed that the removal of the RFTS from the enzymatic pocket of DNMT1, by UHRF1, is involved in its progressive methylation property [114]. When the DNMT1/UHRF1 complex progresses along the DNA, without meeting hemi-methylated DNA, the SRA domain may not interact with the RFTS domain of DNMT1 and, in this way, DNMT1 has no enzymatic activity, avoiding undesired or unfaithful DNA methylation. However, UHRF1 may still interact with DNMT1, but through other domains previously identified [24], and this state would be a kind of standby. An interesting hypothesis would be that this standby also occurs during development, during which a faithful inheritance of DNA methylation patterns would be a brake for cell differentiation or would be an undesired event [67].

3. Concluding Remarks

In conclusion, we propose that both models are correct and compatible, but the events they are based on may be separated in time, with model A most probably happening first and model B happening just after. The following chronological model of the dialogue between UHRF1 and DNMT1 to ensure faithful inheritance of DNA methylation patterns is, therefore, proposed: hemi-methylated DNA–SRA interaction; binding of the TTD to H3K9me3; spacer facilitation of ubiquitination by the RING domain of a dual combination of lysines 14, 18, and 23 of histone H3, and probably also UHRF1 auto-ubiquitination; recruitment (or preferred term interaction) of DNMT1 with mono-ubiquitinated histone H3 and/or ubiquitinated UHRF1; interaction between the SRA domain and/or UBL domain with DNMT1 (via the RFTS domain) and alleviation of the auto-inhibition of DNMT1; and, finally, methylation of the newly synthetized DNA by DNMT1 (Figure 2a,b). This model, however, is based on experimental data obtained with purified proteins, which is a situation that never occurs in vivo as evidenced by numerous studies, reporting that UHRF1 belongs to a macro-molecular complex containing a high number of partners [25,105,120]. This large macro-molecular complex is closely located to the replication fork as it associates with PCNA [18,19,83,121]. We suggest that the UHRF1/DNMT1 tandem slides along the newly synthesized DNA, immediately following the DNA replication machinery in a perpetual complex interacting mode, governed by the status of hemi-methylation of the DNA. The two models of the inheritance of DNA methylation patterns are both likely to ensure faithful DNA methylation duplication. We do not exclude that the two models can act as a kind of double check or double lock to secure fidelity of transmission of methylated profiles. Technological challenges are present that must be overcome to unravel the complexity of the dialogue (direct and/or indirect communication) between UHRF1 and DNMT1. Investigations of in situ domain–domain interactions between the two proteins will help understand when and how each model occurs. This is the price to pay to progress in the understanding of the transmission of this important epigenetic mark.

Author Contributions: C.B., M.A., A.H. and M.M. equally contributed to the writing of this review. C.B. and M.M. designed the figures.

Funding: This work was supported by institutional funds from CNRS, INSERM, Université de Strasbourg (UDS), the Labex INRT, and by grants from, INCA (INCa_4496), INCA (INCa_4454), ANR (EpivarZ, contract n° ANR-16-CE12-0013, La Ligue Nationale contre le Cancer Equipe labellisée (A.H.). Our work was supported by the Agence National de la Recherche (ANR Fluometadn), the Fondation pour la Recherche Médicale (FRM DCM20111223038), Ligue contre le Cancer and by the grant ANR-10-LABX-0030-INRT, and a French State fund managed by the Agence Nationale de la Recherche under the frame program Investissements d'Avenir ANR-10-IDEX-0002-02.

Conflicts of Interest: The authors declare no conflicts of interest.

References

1. Bird, A. DNA methylation patterns and epigenetic memory. *Genes Dev.* **2002**, *16*, 6–21. [CrossRef] [PubMed]
2. Schubeler, D. Function and information content of DNA methylation. *Nature* **2015**, *517*, 321–326. [CrossRef] [PubMed]
3. Jang, H.S.; Shin, W.J.; Lee, J.E.; Do, J.T. Cpg and non-cpg methylation in epigenetic gene regulation and brain function. *Genes* **2017**, *8*.
4. Allis, C.D.; Jenuwein, T. The molecular hallmarks of epigenetic control. *Nat. Rev. Genet.* **2016**, *17*, 487–500. [CrossRef] [PubMed]
5. Portela, A.; Esteller, M. Epigenetic modifications and human disease. *Nat. Biotechnol.* **2010**, *28*, 1057–1068. [CrossRef] [PubMed]
6. Biswas, S.; Rao, C.M. Epigenetics in cancer: Fundamentals and beyond. *Pharmacol. Ther.* **2017**, *173*, 118–134. [CrossRef] [PubMed]
7. Edwards, J.R.; Yarychkivska, O.; Boulard, M.; Bestor, T.H. DNA methylation and DNA methyltransferases. *Epigenetics Chromatin* **2017**, *10*, 23. [CrossRef]
8. Aran, D.; Toperoff, G.; Rosenberg, M.; Hellman, A. Replication timing-related and gene body-specific methylation of active human genes. *Hum. Mol. Genet.* **2011**, *20*, 670–680. [CrossRef]
9. Aran, D.; Sabato, S.; Hellman, A. DNA methylation of distal regulatory sites characterizes dysregulation of cancer genes. *Genome Biol.* **2013**, *14*, R21. [CrossRef]
10. Yang, X.; Han, H.; De Carvalho, D.D.; Lay, F.D.; Jones, P.A.; Liang, G. Gene body methylation can alter gene expression and is a therapeutic target in cancer. *Cancer Cell* **2014**, *26*, 577–590. [CrossRef]

11. Jurkowska, R.Z.; Ceccaldi, A.; Zhang, Y.; Arimondo, P.B.; Jeltsch, A. DNA methyltransferase assays. *Methods Mol. Biol.* **2011**, *791*, 157–177. [PubMed]
12. Merlo, A.; Herman, J.G.; Mao, L.; Lee, D.J.; Gabrielson, E.; Burger, P.C.; Baylin, S.B.; Sidransky, D. 5′ cpg island methylation is associated with transcriptional silencing of the tumour suppressor p16/cdkn2/mts1 in human cancers. *Nat. Med.* **1995**, *1*, 686–692. [CrossRef] [PubMed]
13. Pei, J.H.; Luo, S.Q.; Zhong, Y.; Chen, J.H.; Xiao, H.W.; Hu, W.X. The association between non-hodgkin lymphoma and methylation of p73. *Tumour Biol. J. Int. Soc. Oncodev. Biol. Med.* **2011**, *32*, 1133–1138. [CrossRef] [PubMed]
14. Stefansson, O.A.; Jonasson, J.G.; Olafsdottir, K.; Hilmarsdottir, H.; Olafsdottir, G.; Esteller, M.; Johannsson, O.T.; Eyfjord, J.E. CpG island hypermethylation of BRCA1 and loss of prb as co-occurring events in basal/triple-negative breast cancer. *Epigenetics* **2011**, *6*, 638–649. [CrossRef]
15. Guan, Z.; Zhang, J.; Song, S.; Dai, D. Promoter methylation and expression of TIMP3 gene in gastric cancer. *Diagn. Pathol.* **2013**, *8*, 110. [CrossRef] [PubMed]
16. Rodriguez, J.; Frigola, J.; Vendrell, E.; Risques, R.A.; Fraga, M.F.; Morales, C.; Moreno, V.; Esteller, M.; Capella, G.; Ribas, M.; et al. Chromosomal instability correlates with genome-wide DNA demethylation in human primary colorectal cancers. *Cancer Res.* **2006**, *66*, 8462–9468. [CrossRef] [PubMed]
17. Dawson, M.A.; Kouzarides, T. Cancer epigenetics: From mechanism to therapy. *Cell* **2012**, *150*, 12–27. [CrossRef] [PubMed]
18. Hervouet, E.; Lalier, L.; Debien, E.; Cheray, M.; Geairon, A.; Rogniaux, H.; Loussouarn, D.; Martin, S.A.; Vallette, F.M.; Cartron, P.F. Disruption of Dnmt1/PCNA/UHRF1 interactions promotes tumorigenesis from human and mice glial cells. *PLoS ONE* **2010**, *5*, e11333. [CrossRef]
19. Pacaud, R.; Brocard, E.; Lalier, L.; Hervouet, E.; Vallette, F.M.; Cartron, P.F. The Dnmt1/PCNA/UHRF1 disruption induces tumorigenesis characterized by similar genetic and epigenetic signatures. *Sci. Rep.* **2014**, *4*, 4230. [CrossRef] [PubMed]
20. Nestor, C.E.; Ottaviano, R.; Reinhardt, D.; Cruickshanks, H.A.; Mjoseng, H.K.; McPherson, R.C.; Lentini, A.; Thomson, J.P.; Dunican, D.S.; Pennings, S.; et al. Rapid reprogramming of epigenetic and transcriptional profiles in mammalian culture systems. *Genome Biol.* **2015**, *16*, 11. [CrossRef]
21. Hopfner, R.; Mousli, M.; Garnier, J.M.; Redon, R.; du Manoir, S.; Chatton, B.; Ghyselinck, N.; Oudet, P.; Bronner, C. Genomic structure and chromosomal mapping of the gene coding for ICBP90, a protein involved in the regulation of the topoisomerase iialpha gene expression. *Gene* **2001**, *266*, 15–23. [CrossRef]
22. Hopfner, R.; Mousli, M.; Jeltsch, J.M.; Voulgaris, A.; Lutz, Y.; Marin, C.; Bellocq, J.P.; Oudet, P.; Bronner, C. ICBP90, a novel human ccaat binding protein, involved in the regulation of topoisomerase IIα expression. *Cancer Res.* **2000**, *60*, 121–128. [PubMed]
23. Fagerberg, L.; Hallstrom, B.M.; Oksvold, P.; Kampf, C.; Djureinovic, D.; Odeberg, J.; Habuka, M.; Tahmasebpoor, S.; Danielsson, A.; Edlund, K.; et al. Analysis of the human tissue-specific expression by genome-wide integration of transcriptomics and antibody-based proteomics. *Mol. Cell. Proteom. MCP* **2014**, *13*, 397–406. [CrossRef] [PubMed]
24. Bostick, M.; Kim, J.K.; Esteve, P.O.; Clark, A.; Pradhan, S.; Jacobsen, S.E. UHRF1 plays a role in maintaining DNA methylation in mammalian cells. *Science* **2007**, *317*, 1760–1764. [CrossRef]
25. Sharif, J.; Muto, M.; Takebayashi, S.; Suetake, I.; Iwamatsu, A.; Endo, T.A.; Shinga, J.; Mizutani-Koseki, Y.; Toyoda, T.; Okamura, K.; et al. The SRA protein Np95 mediates epigenetic inheritance by recruiting Dnmt1 to methylated DNA. *Nature* **2007**, *450*, 908–912. [CrossRef] [PubMed]
26. Li, E.; Bestor, T.H.; Jaenisch, R. Targeted mutation of the DNA methyltransferase gene results in embryonic lethality. *Cell* **1992**, *69*, 915–926. [CrossRef]
27. Muto, M.; Kanari, Y.; Kubo, E.; Takabe, T.; Kurihara, T.; Fujimori, A.; Tatsumi, K. Targeted disruption of *Np95* gene renders murine embryonic stem cells hypersensitive to DNA damaging agents and DNA replication blocks. *J. Biol. Chem.* **2002**, *277*, 34549–34555. [CrossRef]
28. Hermann, A.; Gowher, H.; Jeltsch, A. Biochemistry and biology of mammalian DNA methyltransferases. *Cell. Mol. Life Sci. CMLS* **2004**, *61*, 2571–2587. [CrossRef]
29. Alhosin, M.; Sharif, T.; Mousli, M.; Etienne-Selloum, N.; Fuhrmann, G.; Schini-Kerth, V.B.; Bronner, C. Down-regulation of UHRF1, associated with re-expression of tumor suppressor genes, is a common feature of natural compounds exhibiting anti-cancer properties. *J. Exp. Clin. Cancer Res.* **2011**, *30*, 41. [CrossRef]

30. Mohan, K.N.; Chaillet, J.R. Cell and molecular biology of DNA methyltransferase 1. *Int. Rev. Cell Mol. Biol.* **2013**, *306*, 1–42.
31. Jurkowska, R.Z.; Jeltsch, A. Enzymology of mammalian DNA methyltransferases. *Adv. Exp. Med. Biol.* **2016**, *945*, 87–122. [PubMed]
32. Choudhry, H.; Zamzami, M.A.; Omran, Z.; Wu, W.; Mousli, M.; Bronner, C.; Alhosin, M. Targeting microRNA/UHRF1 pathways as a novel strategy for cancer therapy. *Oncol. Lett.* **2018**, *15*, 3–10. [CrossRef] [PubMed]
33. Gowher, H.; Jeltsch, A. Mammalian DNA methyltransferases: New discoveries and open questions. *Biochem. Soc. Trans.* **2018**, *46*, 1191–1202. [CrossRef] [PubMed]
34. Ashraf, W.; Ibrahim, A.; Alhosin, M.; Zaayter, L.; Ouararhni, K.; Papin, C.; Ahmad, T.; Hamiche, A.; Mely, Y.; Bronner, C.; et al. The epigenetic integrator uhrf1: On the road to become a universal biomarker for cancer. *Oncotarget* **2017**, *8*, 51946–51962. [CrossRef] [PubMed]
35. Alhosin, M.; Omran, Z.; Zamzami, M.A.; Al-Malki, A.L.; Choudhry, H.; Mousli, M.; Bronner, C. Signalling pathways in UHRF1-dependent regulation of tumor suppressor genes in cancer. *J. Exp. Clin. Cancer Res.* **2016**, *35*, 174. [CrossRef] [PubMed]
36. Bronner, C.; Krifa, M.; Mousli, M. Increasing role of UHRF1 in the reading and inheritance of the epigenetic code as well as in tumorogenesis. *Biochem Pharm.* **2013**, *86*, 1643–1649. [CrossRef]
37. Bronner, C. Control of dnmt1 abundance in epigenetic inheritance by acetylation, ubiquitylation, and the histone code. *Sci. Signal.* **2011**, *4*, pe3. [CrossRef]
38. Sidhu, H.; Capalash, N. Uhrf1: The key regulator of epigenetics and molecular target for cancer therapeutics. *Tumour Biol. J. Int. Soc. Oncodev. Biol. Med.* **2017**, *39*, 1010428317692205. [CrossRef] [PubMed]
39. Beck, A.; Trippel, F.; Wagner, A.; Joppien, S.; Felle, M.; Vokuhl, C.; Schwarzmayr, T.; Strom, T.M.; von Schweinitz, D.; Langst, G.; et al. Overexpression of UHRF1 promotes silencing of tumor suppressor genes and predicts outcome in hepatoblastoma. *Clin. Epigenetics* **2018**, *10*, 27. [CrossRef]
40. Boukhari, A.; Alhosin, M.; Bronner, C.; Sagini, K.; Truchot, C.; Sick, E.; Schini-Kerth, V.B.; Andre, P.; Mely, Y.; Mousli, M.; et al. CD47 activation-induced UHRF1 over-expression is associated with silencing of tumor suppressor gene *p16INK4a* in glioblastoma cells. *Anticancer Res.* **2015**, *35*, 149–157.
41. Krifa, M.; Alhosin, M.; Muller, C.D.; Gies, J.P.; Chekir-Ghedira, L.; Ghedira, K.; Mely, Y.; Bronner, C.; Mousli, M. Limoniastrum guyonianum aqueous gall extract induces apoptosis in human cervical cancer cells involving *p16INK4a* re-expression related to *UHRF1* and *Dnmt1* down-regulation. *J. Exp. Clin. Cancer Res.* **2013**, *32*, 30. [CrossRef] [PubMed]
42. Achour, M.; Mousli, M.; Alhosin, M.; Ibrahim, A.; Peluso, J.; Muller, C.D.; Schini-Kerth, V.B.; Hamiche, A.; Dhe-Paganon, S.; Bronner, C. Epigallocatechin-3-gallate up-regulates tumor suppressor gene expression via a reactive oxygen species-dependent down-regulation of UHRF1. *Biochem. Biophys. Res. Commun.* **2013**, *430*, 208–212. [CrossRef] [PubMed]
43. Sharif, T.; Alhosin, M.; Auger, C.; Minker, C.; Kim, J.H.; Etienne-Selloum, N.; Bories, P.; Gronemeyer, H.; Lobstein, A.; Bronner, C.; et al. Aronia melanocarpa juice induces a redox-sensitive p73-related caspase 3-dependent apoptosis in human leukemia cells. *PLoS ONE* **2012**, *7*, e32526. [CrossRef] [PubMed]
44. Abusnina, A.; Keravis, T.; Yougbare, I.; Bronner, C.; Lugnier, C. Anti-proliferative effect of curcumin on melanoma cells is mediated by pde1a inhibition that regulates the epigenetic integrator UHRF1. *Mol. Nutr. Food Res.* **2011**, *55*, 1677–1689. [CrossRef]
45. Abusnina, A.; Alhosin, M.; Keravis, T.; Muller, C.D.; Fuhrmann, G.; Bronner, C.; Lugnier, C. Down-regulation of cyclic nucleotide phosphodiesterase PDE1A is the key event of p73 and UHRF1 deregulation in thymoquinone-induced acute lymphoblastic leukemia cell apoptosis. *Cell. Signal.* **2011**, *23*, 152–160. [CrossRef]
46. Zhang, Y.; Huang, Z.; Zhu, Z.; Zheng, X.; Liu, J.; Han, Z.; Ma, X.; Zhang, Y. Upregulated UHRF1 promotes bladder cancer cell invasion by epigenetic silencing of KiSS1. *PLoS ONE* **2014**, *9*, e104252. [CrossRef]
47. Mousli, M.; Hopfner, R.; Abbady, A.Q.; Monte, D.; Jeanblanc, M.; Oudet, P.; Louis, B.; Bronner, C. ICBP90 belongs to a new family of proteins with an expression that is deregulated in cancer cells. *Br. J. Cancer* **2003**, *89*, 120–127. [CrossRef]
48. Patnaik, D.; Esteve, P.O.; Pradhan, S. Targeting the set and ring-associated (SRA) domain of ubiquitin-like, phd and ring finger-containing 1 (UHRF1) for anti-cancer drug development. *Oncotarget* **2018**, *9*, 26243–26258. [CrossRef]

49. Jia, Y.; Li, P.; Fang, L.; Zhu, H.; Xu, L.; Cheng, H.; Zhang, J.; Li, F.; Feng, Y.; Li, Y.; et al. Negative regulation of Dnmt3a de novo DNA methylation by frequently overexpressed uhrf family proteins as a mechanism for widespread DNA hypomethylation in cancer. *Cell Discov.* **2016**, *2*, 16007. [CrossRef]
50. Chuang, L.S.; Ian, H.I.; Koh, T.W.; Ng, H.H.; Xu, G.; Li, B.F. Human DNA-(cytosine-5) methyltransferase-PCNA complex as a target for p21WAF1. *Science* **1997**, *277*, 1996–2000. [CrossRef]
51. Leonhardt, H.; Page, A.W.; Weier, H.U.; Bestor, T.H. A targeting sequence directs DNA methyltransferase to sites of DNA replication in mammalian nuclei. *Cell* **1992**, *71*, 865–873. [CrossRef]
52. Spada, F.; Haemmer, A.; Kuch, D.; Rothbauer, U.; Schermelleh, L.; Kremmer, E.; Carell, T.; Langst, G.; Leonhardt, H. Dnmt1 but not its interaction with the replication machinery is required for maintenance of DNA methylation in human cells. *J. Cell Biol.* **2007**, *176*, 565–571. [CrossRef] [PubMed]
53. Schermelleh, L.; Haemmer, A.; Spada, F.; Rosing, N.; Meilinger, D.; Rothbauer, U.; Cardoso, M.C.; Leonhardt, H. Dynamics of Dnmt1 interaction with the replication machinery and its role in postreplicative maintenance of DNA methylation. *Nucleic Acids Res.* **2007**, *35*, 4301–4312. [CrossRef] [PubMed]
54. Mudbhary, R.; Hoshida, Y.; Chernyavskaya, Y.; Jacob, V.; Villanueva, A.; Fiel, M.I.; Chen, X.; Kojima, K.; Thung, S.; Bronson, R.T.; et al. UHRF1 overexpression drives DNA hypomethylation and hepatocellular carcinoma. *Cancer Cell* **2014**, *25*, 196–209. [CrossRef]
55. Fatemi, M.; Hermann, A.; Gowher, H.; Jeltsch, A. Dnmt3a and dnmt1 functionally cooperate during de novo methylation of DNA. *Eur. J. Biochem.* **2002**, *269*, 4981–4984. [CrossRef] [PubMed]
56. Jair, K.W.; Bachman, K.E.; Suzuki, H.; Ting, A.H.; Rhee, I.; Yen, R.W.; Baylin, S.B.; Schuebel, K.E. De novo cpg island methylation in human cancer cells. *Cancer Res.* **2006**, *66*, 682–692. [CrossRef] [PubMed]
57. Kilin, V.; Gavvala, K.; Barthes, N.P.; Michel, B.Y.; Shin, D.; Boudier, C.; Mauffret, O.; Yashchuk, V.; Mousli, M.; Ruff, M.; et al. Dynamics of methylated cytosine flipping by UHRF1. *J. Am. Chem. Soc.* **2017**, *139*, 2520–2528. [CrossRef]
58. Greiner, V.J.; Kovalenko, L.; Humbert, N.; Richert, L.; Birck, C.; Ruff, M.; Zaporozhets, O.A.; Dhe-Paganon, S.; Bronner, C.; Mely, Y. Site-selective monitoring of the interaction of the sra domain of UHRF1 with target DNA sequences labeled with 2-aminopurine. *Biochemistry* **2015**, *54*, 6012–6020. [CrossRef]
59. Bronner, C.; Fuhrmann, G.; Chedin, F.L.; Macaluso, M.; Dhe-Paganon, S. UHRF1 links the histone code and DNA methylation to ensure faithful epigenetic memory inheritance. *Genet. Epigenetics* **2010**, *2009*, 29–36. [CrossRef]
60. Achour, M.; Jacq, X.; Ronde, P.; Alhosin, M.; Charlot, C.; Chataigneau, T.; Jeanblanc, M.; Macaluso, M.; Giordano, A.; Hughes, A.D.; et al. The interaction of the sra domain of ICBP90 with a novel domain of DNMT1 is involved in the regulation of *VEGF* gene expression. *Oncogene* **2008**, *27*, 2187–2197. [CrossRef]
61. Avvakumov, G.V.; Walker, J.R.; Xue, S.; Li, Y.; Duan, S.; Bronner, C.; Arrowsmith, C.H.; Dhe-Paganon, S. Structural basis for recognition of hemi-methylated DNA by the SRA domain of human uhrf1. *Nature* **2008**, *455*, 822–825. [CrossRef] [PubMed]
62. Ren, R.; Horton, J.R.; Zhang, X.; Blumenthal, R.M.; Cheng, X. Detecting and interpreting DNA methylation marks. *Curr. Opin. Struct. Biol.* **2018**, *53*, 88–99. [CrossRef] [PubMed]
63. Tauber, M.; Fischle, W. Conserved linker regions and their regulation determine multiple chromatin-binding modes of UHRF1. *Nucleus* **2015**, *6*, 123–132. [CrossRef] [PubMed]
64. Hashimoto, H.; Horton, J.R.; Zhang, X.; Bostick, M.; Jacobsen, S.E.; Cheng, X. The SRA domain of UHRF1 flips 5-methylcytosine out of the DNA helix. *Nature* **2008**, *455*, 826–829. [CrossRef] [PubMed]
65. Smets, M.; Link, S.; Wolf, P.; Schneider, K.; Solis, V.; Ryan, J.; Meilinger, D.; Qin, W.; Leonhardt, H. DNMT1 mutations found in HSANIE patients affect interaction with UHRF1 and neuronal differentiation. *Hum. Mol. Genet.* **2017**, *26*, 1522–1534. [CrossRef] [PubMed]
66. Jung, H.J.; Byun, H.O.; Jee, B.A.; Min, S.; Jeoun, U.W.; Lee, Y.K.; Seo, Y.; Woo, H.G.; Yoon, G. The ubiquitin-like with phd and ring finger domains 1 (UHRF1)/DNA methyltransferase 1 (DNMT1) axis is a primary regulator of cell senescence. *J. Biol. Chem.* **2017**, *292*, 3729–3739. [CrossRef] [PubMed]
67. Sen, G.L.; Reuter, J.A.; Webster, D.E.; Zhu, L.; Khavari, P.A. DNMT1 maintains progenitor function in self-renewing somatic tissue. *Nature* **2010**, *463*, 563–567. [CrossRef]
68. Blanchart, A.; Navis, A.C.; Assaife-Lopes, N.; Usoskin, D.; Aranda, S.; Sontheimer, J.; Ernfors, P. UHRF1 licensed self-renewal of active adult neural stem cells. *Stem Cells* **2018**. [CrossRef]

69. Zhao, J.; Chen, X.; Song, G.; Zhang, J.; Liu, H.; Liu, X. Uhrf1 controls the self-renewal versus differentiation of hematopoietic stem cells by epigenetically regulating the cell-division modes. *Proc. Natl. Acad. Sci. USA* **2017**, *114*, E142–E151. [CrossRef]
70. Murao, N.; Matsubara, S.; Matsuda, T.; Noguchi, H.; Mutoh, T.; Mutoh, M.; Koseki, H.; Namihira, M.; Nakashima, K. Np95/UHRF1 regulates tumor suppressor gene expression of neural stem/precursor cells, contributing to neurogenesis in the adult mouse brain. *Neurosci. Res.* **2018**. [CrossRef]
71. Chen, C.; Zhai, S.; Zhang, L.; Chen, J.; Long, X.; Qin, J.; Li, J.; Huo, R.; Wang, X. Uhrf1 regulates germinal center B cell expansion and affinity maturation to control viral infection. *J. Exp. Med.* **2018**, *215*, 1437–1448. [CrossRef] [PubMed]
72. Obata, Y.; Furusawa, Y.; Endo, T.A.; Sharif, J.; Takahashi, D.; Atarashi, K.; Nakayama, M.; Onawa, S.; Fujimura, Y.; Takahashi, M.; et al. The epigenetic regulator uhrf1 facilitates the proliferation and maturation of colonic regulatory t cells. *Nat. Immunol.* **2014**, *15*, 571–579. [CrossRef] [PubMed]
73. Elia, L.; Kunderfranco, P.; Carullo, P.; Vacchiano, M.; Farina, F.M.; Hall, I.F.; Mantero, S.; Panico, C.; Papait, R.; Condorelli, G.; et al. UHRF1 epigenetically orchestrates smooth muscle cell plasticity in arterial disease. *J. Clin. Investig.* **2018**, *128*, 2473–2486. [CrossRef] [PubMed]
74. Milagre, I.; Stubbs, T.M.; King, M.R.; Spindel, J.; Santos, F.; Krueger, F.; Bachman, M.; Segonds-Pichon, A.; Balasubramanian, S.; Andrews, S.R.; et al. Gender differences in global but not targeted demethylation in IPSC reprogramming. *Cell Rep.* **2017**, *18*, 1079–1089. [CrossRef] [PubMed]
75. Yamashita, M.; Inoue, K.; Saeki, N.; Ideta-Otsuka, M.; Yanagihara, Y.; Sawada, Y.; Sakakibara, I.; Lee, J.; Ichikawa, K.; Kamei, Y.; et al. UHRF1 is indispensable for normal limb growth by regulating chondrocyte differentiation through specific gene expression. *Development* **2018**, *145*, dev157412. [CrossRef] [PubMed]
76. Tittle, R.K.; Sze, R.; Ng, A.; Nuckels, R.J.; Swartz, M.E.; Anderson, R.M.; Bosch, J.; Stainier, D.Y.; Eberhart, J.K.; Gross, J.M. UHRF1 and DNMT1 are required for development and maintenance of the zebrafish lens. *Dev. Biol.* **2011**, *350*, 50–63. [CrossRef] [PubMed]
77. Lydon-Rochelle, M.T.; Cardenas, V.; Nelson, J.L.; Tomashek, K.M.; Mueller, B.A.; Easterling, T.R. Validity of maternal and perinatal risk factors reported on fetal death certificates. *Am. J. Public Health* **2005**, *95*, 1948–1951. [CrossRef] [PubMed]
78. Nady, N.; Lemak, A.; Walker, J.R.; Avvakumov, G.V.; Kareta, M.S.; Achour, M.; Xue, S.; Duan, S.; Allali-Hassani, A.; Zuo, X.; et al. Recognition of multivalent histone states associated with heterochromatin by UHRF1 protein. *J. Biol. Chem.* **2011**, *286*, 24300–24311. [CrossRef] [PubMed]
79. Xie, S.; Jakoncic, J.; Qian, C. UHRF1 double tudor domain and the adjacent phd finger act together to recognize k9me3-containing histone H3 tail. *J. Mol. Biol.* **2012**, *415*, 318–328. [CrossRef]
80. Du, J.; Johnson, L.M.; Jacobsen, S.E.; Patel, D.J. DNA methylation pathways and their crosstalk with histone methylation. *Nat. Rev. Mol. Cell Biol.* **2015**, *16*, 519–532. [CrossRef]
81. Rothbart, S.B.; Krajewski, K.; Nady, N.; Tempel, W.; Xue, S.; Badeaux, A.I.; Barsyte-Lovejoy, D.; Martinez, J.Y.; Bedford, M.T.; Fuchs, S.M.; et al. Association of UHRF1 with methylated H3K9 directs the maintenance of DNA methylation. *Nat. Struct. Mol. Biol.* **2012**, *19*, 1155–1160. [CrossRef] [PubMed]
82. Rothbart, S.B.; Dickson, B.M.; Ong, M.S.; Krajewski, K.; Houliston, S.; Kireev, D.B.; Arrowsmith, C.H.; Strahl, B.D. Multivalent histone engagement by the linked tandem tudor and phd domains of UHRF1 is required for the epigenetic inheritance of DNA methylation. *Genes Dev.* **2013**, *27*, 1288–1298. [CrossRef] [PubMed]
83. Liu, X.; Gao, Q.; Li, P.; Zhao, Q.; Zhang, J.; Li, J.; Koseki, H.; Wong, J. UHRF1 targets DNMT1 for DNA methylation through cooperative binding of hemi-methylated DNA and methylated h3k9. *Nat. Commun.* **2013**, *4*, 1563. [CrossRef] [PubMed]
84. Rajakumara, E.; Wang, Z.; Ma, H.; Hu, L.; Chen, H.; Lin, Y.; Guo, R.; Wu, F.; Li, H.; Lan, F.; et al. PHD finger recognition of unmodified histone H3R2 links UHRF1 to regulation of euchromatic gene expression. *Mol. Cell* **2011**, *43*, 275–284. [CrossRef] [PubMed]
85. Lallous, N.; Legrand, P.; McEwen, A.G.; Ramon-Maiques, S.; Samama, J.P.; Birck, C. The PHD finger of human UHRF1 reveals a new subgroup of unmethylated histone H3 tail readers. *PLoS ONE* **2011**, *6*, e27599. [CrossRef] [PubMed]
86. Hu, L.; Li, Z.; Wang, P.; Lin, Y.; Xu, Y. Crystal structure of PHD domain of UHRF1 and insights into recognition of unmodified histone H3 arginine residue 2. *Cell Res.* **2011**, *21*, 1374–1378. [CrossRef]

87. Wang, C.; Shen, J.; Yang, Z.; Chen, P.; Zhao, B.; Hu, W.; Lan, W.; Tong, X.; Wu, H.; Li, G.; et al. Structural basis for site-specific reading of unmodified R2 of histone H3 tail by UHRF1 PHD finger. *Cell Res.* **2011**, *21*, 1379–1382. [CrossRef]
88. Qin, W.; Wolf, P.; Liu, N.; Link, S.; Smets, M.; La Mastra, F.; Forne, I.; Pichler, G.; Horl, D.; Fellinger, K.; et al. DNA methylation requires a DNMT1 ubiquitin interacting motif (UIM) and histone ubiquitination. *Cell Res.* **2015**, *25*, 911–929. [CrossRef]
89. Ronau, J.A.; Beckmann, J.F.; Hochstrasser, M. Substrate specificity of the ubiquitin and Ubl proteases. *Cell Res.* **2016**, *26*, 441–456. [CrossRef]
90. DaRosa, P.A.; Harrison, J.S.; Zelter, A.; Davis, T.N.; Brzovic, P.; Kuhlman, B.; Klevit, R.E. A bifunctional role for the UHRF1 Ubl domain in the control of hemi-methylated DNA-dependent histone ubiquitylation. *Mol. Cell* **2018**, *72*, 753.e6–765.e6. [CrossRef]
91. Foster, B.M.; Stolz, P.; Mulholland, C.B.; Montoya, A.; Kramer, H.; Bultmann, S.; Bartke, T. Critical role of the UBL domain in stimulating the E3 ubiquitin ligase activity of uhrf1 toward chromatin. *Mol. Cell* **2018**, *72*, 739.e9–752.e9. [CrossRef] [PubMed]
92. Leonhardt, H.; Cardoso, M.C. DNA methylation, nuclear structure, gene expression and cancer. *J. Cell. Biochem. Suppl.* **2000**, *79* (Suppl. 35), 78–83. [CrossRef]
93. Qin, W.; Leonhardt, H.; Pichler, G. Regulation of DNA methyltransferase 1 by interactions and modifications. *Nucleus* **2011**, *2*, 392–402. [CrossRef] [PubMed]
94. Jeltsch, A.; Jurkowska, R.Z. New concepts in DNA methylation. *Trends Biochem. Sci.* **2014**, *39*, 310–318. [CrossRef] [PubMed]
95. Jeltsch, A.; Jurkowska, R.Z. Allosteric control of mammalian DNA methyltransferases—A new regulatory paradigm. *Nucleic Acids Res.* **2016**, *44*, 8556–8575. [CrossRef] [PubMed]
96. Zhang, Z.M.; Liu, S.; Lin, K.; Luo, Y.; Perry, J.J.; Wang, Y.; Song, J. Crystal structure of human DNA methyltransferase 1. *J. Mol. Biol.* **2015**, *427*, 2520–2531. [CrossRef] [PubMed]
97. Fang, J.; Cheng, J.; Wang, J.; Zhang, Q.; Liu, M.; Gong, R.; Wang, P.; Zhang, X.; Feng, Y.; Lan, W.; et al. Hemi-methylated DNA opens a closed conformation of uhrf1 to facilitate its histone recognition. *Nat. Commun.* **2016**, *7*, 11197. [CrossRef] [PubMed]
98. Gelato, K.A.; Tauber, M.; Ong, M.S.; Winter, S.; Hiragami-Hamada, K.; Sindlinger, J.; Lemak, A.; Bultsma, Y.; Houliston, S.; Schwarzer, D.; et al. Accessibility of different histone H3-binding domains of UHRF1 is allosterically regulated by phosphatidylinositol 5-phosphate. *Mol. Cell* **2014**, *54*, 905–919. [CrossRef]
99. Gao, L.; Tan, X.F.; Zhang, S.; Wu, T.; Zhang, Z.M.; Ai, H.W.; Song, J. An intramolecular interaction of UHRF1 reveals dual control for its histone association. *Structure* **2018**, *26*, 304.e3–311.e3. [CrossRef] [PubMed]
100. Misaki, T.; Yamaguchi, L.; Sun, J.; Orii, M.; Nishiyama, A.; Nakanishi, M. The replication foci targeting sequence (RFTS) of DNMT1 functions as a potent histone H3 binding domain regulated by autoinhibition. *Biochem. Biophys. Res. Commun.* **2016**, *470*, 741–747. [CrossRef]
101. Nishiyama, A.; Yamaguchi, L.; Sharif, J.; Johmura, Y.; Kawamura, T.; Nakanishi, K.; Shimamura, S.; Arita, K.; Kodama, T.; Ishikawa, F.; et al. Uhrf1-dependent H3K23 ubiquitylation couples maintenance DNA methylation and replication. *Nature* **2013**, *502*, 249–253. [CrossRef] [PubMed]
102. Ishiyama, S.; Nishiyama, A.; Saeki, Y.; Moritsugu, K.; Morimoto, D.; Yamaguchi, L.; Arai, N.; Matsumura, R.; Kawakami, T.; Mishima, Y.; et al. Structure of the dnmt1 reader module complexed with a unique two-mono-ubiquitin mark on histone H3 reveals the basis for DNA methylation maintenance. *Mol. Cell* **2017**, *68*, 350.e357–360.e357. [CrossRef] [PubMed]
103. Harrison, J.S.; Cornett, E.M.; Goldfarb, D.; DaRosa, P.A.; Li, Z.M.; Yan, F.; Dickson, B.M.; Guo, A.H.; Cantu, D.V.; Kaustov, L.; et al. Hemi-methylated DNA regulates DNA methylation inheritance through allosteric activation of H3 ubiquitylation by UHRF1. *eLife* **2016**, *5*, e17101. [CrossRef] [PubMed]
104. Jenkins, Y.; Markovtsov, V.; Lang, W.; Sharma, P.; Pearsall, D.; Warner, J.; Franci, C.; Huang, B.; Huang, J.; Yam, G.C.; et al. Critical role of the ubiquitin ligase activity of UHRF1, a nuclear ring finger protein, in tumor cell growth. *Mol. Biol. Cell* **2005**, *16*, 5621–5629. [CrossRef] [PubMed]
105. Ibrahim, A.; Alhosin, M.; Papin, C.; Ouararhni, K.; Omran, Z.; Zamzami, M.A.; Al-Malki, A.L.; Choudhry, H.; Mely, Y.; Hamiche, A.; et al. Thymoquinone challenges UHRF1 to commit auto-ubiquitination: A key event for apoptosis induction in cancer cells. *Oncotarget* **2018**, *9*, 28599–28611. [CrossRef]

106. Li, T.; Wang, L.; Du, Y.; Xie, S.; Yang, X.; Lian, F.; Zhou, Z.; Qian, C. Structural and mechanistic insights into uhrf1-mediated DNMT1 activation in the maintenance DNA methylation. *Nucleic Acids Res.* **2018**, *46*, 3218–3231. [CrossRef]
107. Komander, D.; Rape, M. The ubiquitin code. *Annu. Rev. Biochem.* **2012**, *81*, 203–229. [CrossRef]
108. Peterson, C.L.; Laniel, M.A. Histones and histone modifications. *Curr. Biol.* **2004**, *14*, R546–R551. [CrossRef]
109. Yamaguchi, L.; Nishiyama, A.; Misaki, T.; Johmura, Y.; Ueda, J.; Arita, K.; Nagao, K.; Obuse, C.; Nakanishi, M. Usp7-dependent histone h3 deubiquitylation regulates maintenance of DNA methylation. *Sci. Rep.* **2017**, *7*, 55. [CrossRef]
110. Felle, M.; Joppien, S.; Nemeth, A.; Diermeier, S.; Thalhammer, V.; Dobner, T.; Kremmer, E.; Kappler, R.; Langst, G. The usp7/dnmt1 complex stimulates the DNA methylation activity of DNMT1 and regulates the stability of UHRF1. *Nucleic Acids Res.* **2011**, *39*, 8355–8365. [CrossRef]
111. Bronner, C.; Achour, M.; Arima, Y.; Chataigneau, T.; Saya, H.; Schini-Kerth, V.B. The UHRF family: Oncogenes that are drugable targets for cancer therapy in the near future? *Pharmacol. Ther.* **2007**, *115*, 419–434. [CrossRef]
112. Takeshita, K.; Suetake, I.; Yamashita, E.; Suga, M.; Narita, H.; Nakagawa, A.; Tajima, S. Structural insight into maintenance methylation by mouse DNA methyltransferase 1 (DNMT1). *Proc. Natl. Acad. Sci. USA* **2011**, *108*, 9055–9059. [CrossRef] [PubMed]
113. Syeda, F.; Fagan, R.L.; Wean, M.; Avvakumov, G.V.; Walker, J.R.; Xue, S.; Dhe-Paganon, S.; Brenner, C. The replication focus targeting sequence (RFTS) domain is a DNA-competitive inhibitor of DNMT1. *J. Biol. Chem.* **2011**, *286*, 15344–15351. [CrossRef] [PubMed]
114. Berkyurek, A.C.; Suetake, I.; Arita, K.; Takeshita, K.; Nakagawa, A.; Shirakawa, M.; Tajima, S. The DNA methyltransferase DNMT1 directly interacts with the set and ring finger-associated (SRA) domain of the multifunctional protein UHRF1 to facilitate accession of the catalytic center to hemi-methylated DNA. *J. Biol. Chem.* **2014**, *289*, 379–386. [CrossRef] [PubMed]
115. Pichler, G.; Wolf, P.; Schmidt, C.S.; Meilinger, D.; Schneider, K.; Frauer, C.; Fellinger, K.; Rottach, A.; Leonhardt, H. Cooperative DNA and histone binding by UHRF2 links the two major repressive epigenetic pathways. *J. Cell. Biochem.* **2011**, *112*, 2585–2593. [CrossRef] [PubMed]
116. Zhang, J.; Gao, Q.; Li, P.; Liu, X.; Jia, Y.; Wu, W.; Li, J.; Dong, S.; Koseki, H.; Wong, J. S phase-dependent interaction with DNMT1 dictates the role of UHRF1 but not UHRF2 in DNA methylation maintenance. *Cell Res.* **2011**, *21*, 1723–1739. [CrossRef] [PubMed]
117. Bashtrykov, P.; Jankevicius, G.; Jurkowska, R.Z.; Ragozin, S.; Jeltsch, A. The uhrf1 protein stimulates the activity and specificity of the maintenance DNA methyltransferase DNMT1 by an allosteric mechanism. *J. Biol. Chem.* **2014**, *289*, 4106–4115. [CrossRef]
118. Vaughan, R.M.; Dickson, B.M.; Whelihan, M.F.; Johnstone, A.L.; Cornett, E.M.; Cheek, M.A.; Ausherman, C.A.; Cowles, M.W.; Sun, Z.W.; Rothbart, S.B. Chromatin structure and its chemical modifications regulate the ubiquitin ligase substrate selectivity of UHRF1. *Proc. Natl. Acad. Sci. USA* **2018**, *115*, 8775–8780. [CrossRef]
119. Arita, K.; Ariyoshi, M.; Tochio, H.; Nakamura, Y.; Shirakawa, M. Recognition of hemi-methylated DNA by the sra protein UHRF1 by a base-flipping mechanism. *Nature* **2008**, *455*, 818–821. [CrossRef] [PubMed]
120. Ferry, L.; Fournier, A.; Tsusaka, T.; Adelmant, G.; Shimazu, T.; Matano, S.; Kirsh, O.; Amouroux, R.; Dohmae, N.; Suzuki, T.; et al. Methylation of DNA ligase 1 by G9A/GLP recruits UHRF1 to replicating DNA and regulates DNA methylation. *Mol. Cell* **2017**, *67*, 550.e5–565.e5. [CrossRef]
121. Miura, M.; Watanabe, H.; Sasaki, T.; Tatsumi, K.; Muto, M. Dynamic changes in subnuclear Np95 location during the cell cycle and its spatial relationship with DNA replication foci. *Exp. Cell Res.* **2001**, *263*, 202–208. [CrossRef] [PubMed]

© 2019 by the authors. Licensee MDPI, Basel, Switzerland. This article is an open access article distributed under the terms and conditions of the Creative Commons Attribution (CC BY) license (http://creativecommons.org/licenses/by/4.0/).

Review

DNA Methylation Reprogramming during Mammalian Development

Yang Zeng [1,2,3] and Taiping Chen [1,2,3,*]

1. Department of Epigenetics and Molecular Carcinogenesis, The University of Texas MD Anderson Cancer Center, 1808 Park Road 1C, Smithville, TX 78957, USA; yzeng1@mdanderson.org
2. Center for Cancer Epigenetics, The University of Texas MD Anderson Cancer Center, Smithville, TX 78957, USA
3. Program in Genetics and Epigenetics, The University of Texas MD Anderson Cancer Center UTHealth Graduate School of Biomedical Sciences, Houston, TX 77030, USA
* Correspondence: tchen2@mdanderson.org; Tel.: +1-512-237-9479; Fax: +1-512-237-2475

Received: 6 March 2019; Accepted: 25 March 2019; Published: 29 March 2019

Abstract: DNA methylation (5-methylcytosine, 5mC) is a major form of DNA modification in the mammalian genome that plays critical roles in chromatin structure and gene expression. In general, DNA methylation is stably maintained in somatic tissues. However, DNA methylation patterns and levels show dynamic changes during development. Specifically, the genome undergoes two waves of global demethylation and remethylation for the purpose of producing the next generation. The first wave occurs in the germline, initiated with the erasure of global methylation in primordial germ cells (PGCs) and completed with the establishment of sex-specific methylation patterns during later stages of germ cell development. The second wave occurs after fertilization, including the erasure of most methylation marks inherited from the gametes and the subsequent establishment of the embryonic methylation pattern. The two waves of DNA methylation reprogramming involve both distinct and shared mechanisms. In this review article, we provide an overview of the key reprogramming events, focusing on the important players in these processes, including DNA methyltransferases (DNMTs) and ten-eleven translocation (TET) family of 5mC dioxygenases.

Keywords: DNA methylation; embryogenesis; germ cells; DNMTs; TETs

---

1. Introduction

DNA Methylation—the addition of a methyl group to the 5-position of cytosine, forming 5-methylcytosine (5mC)—is a major form of DNA modification in many, but not all, eukaryotic organisms. In mammals, DNA methylation mainly occurs in the context of CpG dinucleotides, forming a symmetrical pattern on both strands. In the mouse genome, the majority (~60–80%) of CpG dinucleotides are methylated. Non-CpG (i.e., CpA, CpT, or CpC) methylation is rare except in special cell types, such as embryonic stem cells (ESCs), oocytes, and neurons [1]. 5mC distribution in the genome is bimodal. In general, repetitive sequences, such as transposons and centromeric and pericentric repeats, are heavily methylated, gene bodies of highly expressed genes are also methylated, whereas CpG islands (CGIs), i.e., 500–2000-bp GC-rich sequences that are frequently present in promoter regions, are usually devoid of methylation [2]. DNA methylation is essential for mammalian development and is involved in a variety of biological processes, including transcriptional regulation, transposon silencing, X chromosome inactivation, and genomic imprinting [3]. Aberrant DNA methylation patterns and mutations of genes encoding DNA methylation enzymes or regulators are associated with developmental disorders and cancer [4,5].

As a stable epigenetic mark that is heritable through cell division, DNA methylation is an important component of the cellular memory mechanism that maintains cell identities. However, the epigenome, including DNA methylation, needs to be reprogrammed to a totipotent state for producing the next generation. There are two waves of global demethylation and remethylation during the mammalian life cycle (Figure 1), one occurring during germ cell development and the other occurring during early embryogenesis [6,7]. Epigenetic reprogramming in the germline involves the erasure of somatic methylation patterns in primordial germ cells (PGCs) and subsequent establishment of sex-specific germ cell methylation patterns, including methylation marks in imprinting control regions (ICRs). Epigenetic reprogramming in early embryos involves erasure of most methylation marks inherited from the gametes (exceptions include methylation marks in ICRs and some retrotransposons) at preimplantation stages and reestablishment of global DNA methylation patterns upon implantation. In this review, we discuss the highly dynamic and regulated processes of DNA demethylation and remethylation, focusing on the important enzymes and regulators that are involved in these reprogramming events.

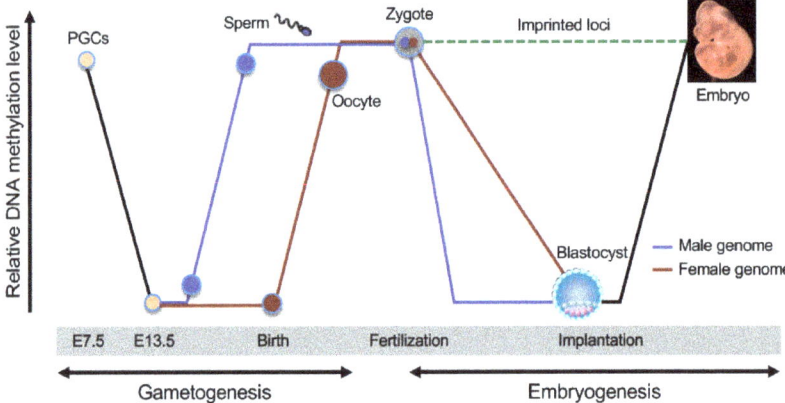

Figure 1. Dynamic changes in DNA methylation during mammalian development. Schematically shown are the two waves of global DNA demethylation and remethylation in the life cycle (adapted from [6]). Primordial germ cells (PGCs) initially have high levels of DNA methylation. Global demethylation occurs during PGC expansion and migration. At later stages of germ cell development (before birth in male and after birth in female), de novo methylation results in the establishment of sex-specific germ cell methylation patterns, including methylation marks at imprinted loci. Shortly after fertilization, the methylation marks inherited from the gametes are erased again (except those at imprinted loci and some retrotransposons), with the paternal genome undergoing active demethylation and the maternal genome undergoing passive demethylation. Upon implantation, a wave of de novo methylation establishes the initial embryonic methylation pattern.

2. DNA Methyltransferases and Regulators

As CpG/CpG dyads are symmetrical, Holliday and Pugh, and Riggs independently proposed that methylated CpG sites could be heritable through semi-conservative DNA replication [8,9]. The theory would predict at least two catalytic activities—de novo methylation activity for the establishment of methylation patterns and maintenance methylation activity for converting hemi-methylated CpG sites to fully methylated ones during DNA replication. Subsequent work identified DNA methyltransferases (DNMTs) with distinct properties and their accessory factors, including ubiquitin−like with plant homeodomain (PHD) and really interesting new gene (RING) finger domains 1 (UHRF1) and DNMT3L [10].

2.1. DNMT1 and UHRF1—Key Components of Maintenance Methylation Machinery

Dnmt1, the first mammalian DNA methyltransferase gene discovered, was cloned from murine cells [11]. It has several transcription start sites, producing three major isoforms [12,13]. *Dnmt1s*, expressed in somatic cells, encodes the full-length DNMT1 protein of 1620 amino acids. *Dnmt1o*, specifically expressed in oocytes, encodes a protein product that lacks the N-terminal 118 amino acids due to translation from a downstream start codon. Compared to DNMT1, DNMT1o has similar catalytic activity but appears to be more stable [14]. *Dnmt1p*, specifically expressed in pachytene spermatocytes, does not produce a protein product. Human DNMT1, with full-length protein consisting of 1616 amino acids, shows ~80% sequence identity to murine DNMT1.

DNMT1 contains a C-terminal catalytic domain that harbors specific motifs (I-X) characteristic of all DNA methyltransferases and a unique N-terminal regulatory domain that harbors several functional domains. These domains include a nuclear localization signal (NLS) that imports DNMT1 to the nucleus, a proliferating cell nuclear antigen (PCNA) binding domain (PBD) that interacts with the DNA replication machinery, a replication foci-targeting sequence (RFTS) that targets DNMT1 to DNA replication foci, a CXXC zinc finger domain that binds unmethylated DNA, and two bromo-adjacent homology (BAH) domains and a glycine-lysine (GK) repeat that are required for the catalytic activity of DNMT1 (Figure 2A). Crystallography data reveal that, in the absence of DNA substrates, the RFTS blocks the catalytic center, suggesting that the N-terminal regulatory region plays a self-inhibitory role for DNMT1 activity [15].

In vitro biochemical assays indicate that DNMT1 prefers DNA substrates containing hemi-methylated CpG sites [16]. *Dnmt1* expression is activated by cell cycle-dependent transcription factors and thus is present at high levels in proliferating cells [17]. Immunofluorescence experiments show enrichment of DNMT1 in replication foci during S phase of the cell cycle, suggesting that its function is coupled with DNA replication [18]. Genetic disruption of *Dnmt1* in mouse ESCs (mESCs) results in global loss of methylation [19,20]. These lines of evidence suggest that DNMT1 plays a major role in maintaining DNA methylation through cell division (Figure 2B). DNMT1 also has de novo methylation activity, which is likely repressed in cells. For example, a recent study suggests that DNMT1-mediated de novo methylation contributes to abnormal hypermethylation in mouse oocytes deficient for the maternal factor STELLA (also known as PGC7 and DPPA3) [21].

UHRF1, also known as NP95 (nuclear protein of 95 kDa) in mouse or ICBP90 (inverted CCAAT box–binding protein of 90 kDa) in human, is an essential accessory factor of DNMT1. *Uhrf1* knockout (KO) mESCs exhibit global hypomethylation [22,23], similar to the phenotype of *Dnmt1* KO mESCs [19,20]. UHRF1 and DNMT1 form a complex and colocalize in replication foci and heterochromatin, and in the absence of UHRF1, DNMT1 fails to localize to these regions [22–24]. These findings suggest that UHRF1 is critical in recruiting DNMT1 to hemi-methylated CpG sites during DNA replication (Figure 2B).

UHRF1 contains five conserved domains, all of which have been implicated in DNMT1 function (Figure 2A). The Su(var)3-9, Enhancer of zeste, and Trithorax (SET)- and RING-associated (SRA) domain preferentially binds hemi-methylated DNA and plays an important role in loading DNMT1 onto newly synthesized DNA [22,23]. The tandem Tudor domain (TTD) and the plant homeodomain (PHD) cooperatively interact with the N-terminal tail of histone H3 by recognizing a specific histone modification signature consisting of trimethylated Lys9 (H3K9me3) and unmethylated Arg2 (H3R2me0) [24–29]. The really interesting new gene (RING) domain has E3 ubiquitin ligase activity and monoubiquitinates several lysine residues in the N-terminal tail of histone H3, creating DNMT1-binding sites [30–32]. There is evidence that SRA domain-mediated recognition of hemi-methylated DNA allosterically activates RING domain-mediated H3 ubiquitination and facilitates histone binding [32,33]. Recently, the ubiquitin-like (UBL) domain was shown to interact with the E2 ubiquitin conjugating enzyme UBE2D to facilitate H3 monoubiquitination [34,35]. Thus, UHRF1 acts as both a 'writer' and 'reader' of histone marks and targets DNMT1 to its DNA substrates via complex, multivalent interactions with chromatin.

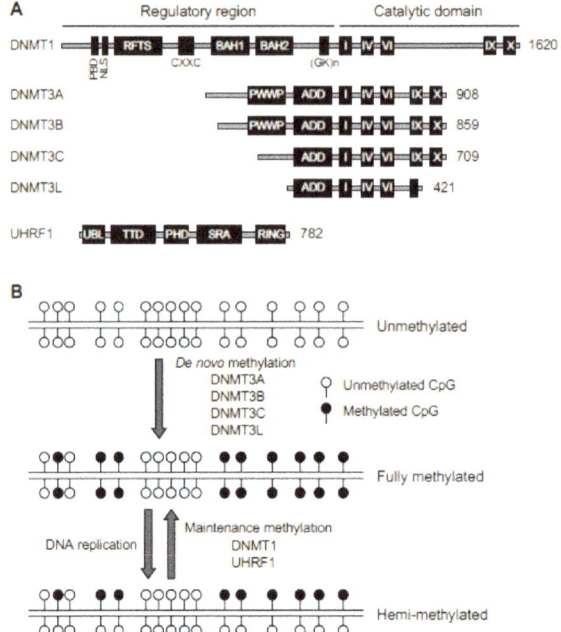

Figure 2. DNA methylation machinery. (**A**) The protein domains in DNA methyltransferases (DNMTs) and ubiquitin−like with plant homeodomain (PHD) and really interesting new gene (RING) finger domains 1 (UHRF1) are shown (the sizes/amino acid numbers refer to mouse proteins). The DNMT1 and DNMT3 families of proteins share conserved catalytic motifs (I–X) in the C-terminal catalytic domains (DNMT3L lacks catalytic activity because some essential motifs are missing or mutated) but have distinct N-terminal regulatory regions. PBD, proliferating cell nuclear antigen (PCNA)-binding domain; NLS, nuclear localization signal; RFTS, replication foci targeting sequence; CXXC, a cysteine-rich zinc finger domain; BAH, bromo-adjacent homology domain; (GK)n, glycine/lysine repeats; PWWP, proline-tryptophan-tryptophan-proline domain; ADD, ATRX-DNMT3-DNMT3L domain; UBL, ubiquitin-like domain; TTD, tandem Tudor domain; PHD, plant homeodomain; SRA, Su(var)3-9, Enhancer of zeste, and Trithorax (SET)- and RING-associated domain; RING, really interesting new gene domain. (**B**) De novo and maintenance methylation activities. The de novo methyltransferases (DNMT3A, DNMT3B, and DNMT3C), in complex with their accessory factor DNMT3L, methylate unmethylated CpG sites to establish methylation patterns. The maintenance methyltransferase DNMT1, in complex with its accessory factor UHRF1, methylates hemi-methylated CpG sites at each round of DNA replication to maintain methylation patterns.

2.2. DNMT3A, DNMT3B, DNMT3C, and DNMT3L—Key Components of de novo Methylation Machinery

By homology searches using bacterial DNA methyltransferase genes, Okano et al. identified two mammalian homologues, *Dnmt3a* and *Dnmt3b* [36]. *Dnmt3a* has two major promoters, producing the *Dnmt3a1* and *Dnmt3a2* isoforms. *Dnmt3a1* encodes the full-length DNMT3A protein (908 amino acids in mouse and 912 amino acids in human), and *Dnmt3a2*, driven by a downstream intronic promoter, encodes a protein product that lacks the N-terminal 219 (mouse) or 223 (human) amino acids. DNMT3A and DNMT3A2 have similar catalytic activities [37]. Due to alternative splicing, *Dnmt3b* produces multiple isoforms, with some encoding active enzymes (e.g., full-length DNMT3B1, 859 amino acids in mouse and 853 amino acids in human) and others encoding catalytically inactive protein products. Some inactive DNMT3B proteins function as regulators of DNA methylation [38,39].

DNMT3A and DNMT3B have similar structures (Figure 2A), including a C-terminal catalytic domain with the conserved catalytic motifs and an N-terminal regulatory domain with no sequence similarity to DNMT1. The N-terminal region contains two chromatin-interaction domains that likely play important roles in determining the functional specificity of DNMT3A and DNMT3B: A proline-tryptophan-tryptophan-proline (PWWP) domain that recognizes H3K36me3 via an aromatic cage and also binds DNA through a positively charged surface [40–44] and an ATRX-DNMT3-DNMT3L (ADD) domain that interacts with the histone H3 tail with H3K4me0 [45]. Crystal structure indicates that the ADD domain of DNMT3A interacts with its own catalytic domain to inhibit enzymatic activity, resulting in autoinhibition. H3K4me0 peptide, but not H3K4me3 peptide, can disrupt the interaction between the ADD and catalytic domains and thus releases the autoinhibition [46], suggesting that histone marks not only create binding sites for DNMT3A/3B but also stimulate their catalytic activity.

Based on several lines of evidence, DNMT3A and DNMT3B primarily carry out de novo DNA methylation (Figure 2B). First, their expression correlates with de novo methylation during development. Specifically, *Dnmt3a* (mainly *Dnmt3a2*) and *Dnmt3b* (mainly *Dnmt3b1*) are highly expressed in early embryos (as well as mESCs) and are downregulated in differentiated cells (mainly *Dnmt3a1*, *Dnmt3b2*, and *Dnmt3b3*), and *Dnmt3a* (mainly *Dnmt3a2*) is also abundantly expressed in developing germ cells [36,47–49]. Second, in vitro assays indicate that DNMT3A and DNMT3B methylate DNA substrates containing unmethylated and hemi-methylated CpG sites with equal efficiency [36]. Third, acute deletion of *Dnmt3a* and *Dnmt3b* in mESCs does not affect global methylation but prevents de novo methylation of newly integrated proviruses [50]. It is worth mentioning that *Dnmt3a/3b* double KO mESCs show gradual loss of methylation with continuous culturing [51], suggesting that, while DNMT1 is the major maintenance enzyme, DNMT3A and DNMT3B are also required for stable and faithful maintenance of DNA methylation.

Dnmt3c, a *Dnmt3b* duplicated gene present only in rodent genomes, was initially identified as a pseudogene for lack of expression and recognizable open reading frame [52]. Recent work showed that *Dnmt3c* is expressed during spermatogenesis, encoding a protein of 709 amino acids that is similar to DNMT3B but lacks the PWWP domain in the N-terminal regulatory region. Genetic evidence demonstrates that *Dnmt3c* is not required for mouse development but is essential for normal spermatogenesis by methylating and silencing special transposons in the male germline [53].

Dnmt3l, another member of the *Dnmt3* family, encodes a catalytically inactive protein that contains some, but not all, catalytic motifs in the C-terminal region and an ADD domain, but not a PWWP domain, in the N-terminal region [54–56]. Crystal structure reveals that DNMT3A and DNMT3L form a tetramer, with two DNMT3A molecules in the middle and DNMT3L on each side [57,58]. DNMT3L substantially stimulates the catalytic activity of DNMT3A and DNMT3B in vitro [59–61]. Recent work suggests that DNMT3L is also involved in maintaining DNMT3A stability [62]. In addition, the ADD domain of DNMT3L binds H3K4me0 and likely plays a role in determining the specificity of DNMT3A-mediated DNA methylation [63]. These results indicate that DNMT3L is an important accessory factor of DNMT3A (and perhaps DNMT3B as well) (Figure 2B). Contrary to previous evidence that DNMT3L is a positive regulator [56,59–61,64–67], Neri et al. reported gain of methylation at bivalent promoters in *Dnmt3l* knockdown (KD) mESCs and concluded that DNMT3L antagonizes DNA methylation at some genomic regions [68]. However, genome-wide analysis of *Dnmt3l*-deficient mESCs revealed only loss of methylation, mostly at DNMT3A target loci, in a recent study [62], disputing the finding and conclusion of the Neri paper [68].

2.3. DNMT2/TRDMT1

Another protein, initially named DNMT2, contains the catalytic motifs characteristic of DNMTs but is not required for either de novo or maintenance DNA methylation [69]. Subsequent work shows that it methylates aspartic acid tRNA at cytosine 38. Thus, it has been renamed tRNA aspartic acid [D] methyltransferase 1 (TRDMT1) [70]. While *Dnmt2* KO mice are viable and grossly normal [69], recent evidence suggests that DNMT2/TRDMT1-dependent RNA modifications play an important role in intergenerational transmission of paternally acquired metabolic disorders to offspring by regulating the sperm small RNA expression profile, including tRNA- and rRNA- derived small RNAs [71].

3. DNA Demethylation

Although DNA methylation is generally stable in somatic tissues, loss of methylation occurs during development, cellular differentiation, aging, and in cancer cells. In general, loss of methylation can be achieved by two mechanisms: DNA replication-dependent passive dilution of 5mC and enzyme-mediated active removal/replacement of 5mC [72].

3.1. Passive DNA Demethylation

During DNA replication, methylation patterns are maintained by the maintenance methylation machinery, including DNMT1 and its accessory factor UHRF1. Therefore, functional deficiency in maintenance methylation can lead to replication-dependent dilution of 5mC, known as passive DNA demethylation. This mechanism, which allows efficient erasure of global DNA methylation, is relevant in multiple biological processes. In preimplantation embryos, erasure of DNA methylation marks inherited from the oocyte (i.e., in the maternal genome) is mainly through passive demethylation during cleavage division, presumably due to the exclusion of DNMT1 from the nuclei [6,7]. In PGCs, the first phase of global demethylation is a passive process, as a result of the silencing of several key DNA methylation enzymes and regulators, including UHRF1 [73]. In mESC culture, there exists a small population of totipotent cells, similar to 2-cell (2C) embryos, at any given time. It is believed that mESCs convert to the transient 2C-like state to recover shortened telomeres and repair DNA damage. Recent work reveals that, due to UHRF1 and DNMT1 degradation, 2C-like cells are severely hypomethylated, which is required for telomere elongation [74]. Cancer cells generally exhibit global loss of methylation and regional gain of methylation. While the mechanisms underlying these aberrant changes are complex, overexpression of PRMT6, an enzyme that deposits asymmetric dimethylation of H3R2 (H3R2me2a), was recently shown to contribute to global hypomethylation in cancer cells, as H3R2me2a inhibits the recruitment of the UHRF1-DNMT1 complex to chromatin [75].

3.2. Active DNA Demethylation

Rapid loss of DNA methylation marks can also occur in slowly or non-dividing cells, which cannot be explained by passive demethylation. As the methyl group in 5mC is linked to cytosine by a carbon-carbon bond, demethylation by direct cleavage of the methyl group is considered infeasible. Active DNA demethylation, which is independent of DNA replication, refers to processes that result in indirect removal of 5mC, involving enzyme-mediated modifications of 5mC followed by replacement with unmodified cytosine. Significant progress has been made in understanding the mechanism of active demethylation since the discovery that 5mC can be converted to oxidized forms by the ten-eleven translocation (TET) enzymes [76,77].

There are three members in the TET family, TET1, TET2, and TET3 (Figure 3A). They are capable of oxidizing 5mC into 5-hydroxymethylcytosine (5hmC), 5-formylcytosine (5fC), and 5-carboxylcytosine (5caC) [76–79]. All three TET proteins have similar structures, with a C-terminal catalytic domain and an N-terminal regulatory domain. The C-terminal catalytic domain contains a cysteine-rich region (Cys) and a double-stranded β-helix (DSBH) domain. The catalytic core region preferentially recognizes 5mC as a substrate and also binds to 5mC oxidized derivatives [80]. TET1 and TET3, but not TET2, contain a DNA-binding CXXC zinc finger domain in the N-terminal regulatory domain. During evolution, the ancestral *Tet2* gene was split into two distinct genes, *Idax* and the current *Tet2*. IDAX, which contains a CXXC domain, interacts with TET2, and may be involved in recruiting TET2 to its genomic targets [81].

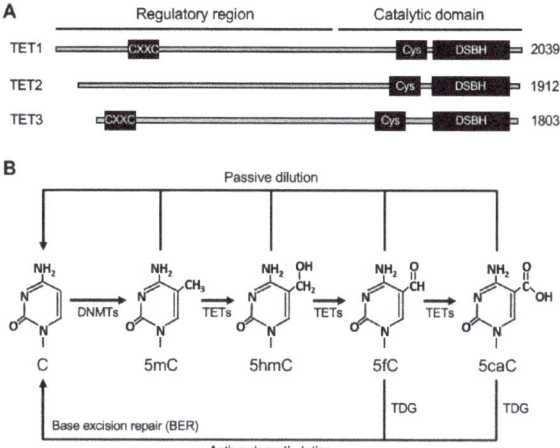

Figure 3. Ten-eleven translocation (TET) proteins and relevant DNA demethylation pathways. (**A**) The protein domains in TET proteins (TET1, TET2 and TET3) are shown (the sizes/amino acid numbers refer to mouse proteins). Their C-terminal catalytic domains contain two characteristic sequence features, a cysteine-rich region (Cys) and a double-stranded β helix (DSBH) fold. Their N-terminal regulatory regions are less conserved, with TET1 and TET3 containing a CXXC zinc finger domain. (**B**) DNA demethylation pathways involving TETs. TET proteins initiate DNA demethylation by oxidizing 5-methylcytosine (5mC) to 5-hydroxymethylcytosine (5hmC), which can be further oxidized to 5-formylcytosine (5fC) and 5-carboxylcytosine (5caC). 5fC and 5caC can be excised by thymine DNA glycosylase (TDG). The residual abasic site can then be repaired by the base excision repair (BER) pathway to complete 'active' demethylation. 5mC, 5hmC, 5fC, and 5caC can also be removed through DNA replication-coupled 'passive' dilution.

Studies over the last decade have demonstrated that 5hmC, 5fC, and 5caC serve as intermediates of DNA demethylation (Figure 3B). 5fC and 5caC can be recognized and excised from DNA by thymine DNA glycosylase (TDG) [79,82,83]. The residual abasic site can then be repaired by the base excision repair (BER) pathway. Thus, this process results in the replacement of 5mC by an unmodified cytosine, achieving 'active' demethylation. It is important to point out that passive dilution is also involved in TET-mediated DNA demethylation. Since 5mC oxidized derivatives are poorly recognized by DNMT1 [84], these marks could be lost owing to the lack of maintenance during DNA replication.

4. DNA Methylation Reprogramming in the Germline

In early postimplantation embryos, a small number of cells are set aside as PGCs that eventually develop into germ cells (sperm or oocyte) for reproduction. Epigenetic reprogramming in the germline involves the erasure of somatic lineage epigenetic marks (including DNA methylation) in PGCs and subsequent establishment of germ cell-specific epigenomes.

4.1. Biphasic Demethylation in PGCs

In mice, PGCs are specified from the posterior epiblasts at the onset of gastrulation around E6.25-7.25, in response to instructive signals from extraembryonic tissues, including the transcription factors BLIMP1, PRDM14, and BMP4 signaling [85]. After specification, PGCs migrate from the proximal epiblast along the hindgut (~E7.75) and reach the genital ridges (~E10.5) and subsequently start sex differentiation (~E12.5).

As the epiblast rapidly adopts somatic epigenetic features after implantation, PGCs are initially very similar to their somatic counterparts in DNA methylation patterns, transcription profiles, and chromatin modifications, including silencing of pluripotency genes and germline-specific genes by DNA methylation [7]. At this point, PGCs display high levels of 5mC and low levels of 5hmC. In the initial stage of migration at E7.5–E8.5, PGCs arrest in the G2 phase of the cell cycle, then they start rapid proliferation until at least E12.5 [7]. Genome-wide DNA methylation profiling reveals that PGCs undergo DNA demethylation in two distinct phases [86,87] (Figure 4). The first phase, starting at ~E8.5, occurs during PGC proliferation and migration, which results in genome-wide loss of methylation, involving almost all genomic sequences. Passive demethylation is predominantly responsible at this phase, due to deficiency in major components of the methylation machinery. For example, *Uhrf1* is repressed and excluded from the nucleus at E9.5. Only after E12.5, *Uhrf1* expression starts to increase and DNMT1 localization is slowly restored back to replication foci [73]. The second phase occurs from E9.5 to E13.5, which leads to demethylation of specific loci, including imprinted genes, CGIs of the inactive X chromosome in females, and germline-specific and meiosis-specific genes [86–89]. Active demethylation is mainly responsible for removing methylation marks at these loci, which are protected from passive demethylation in early PGCs. TET1- and TET2-mediated conversion of 5mC to 5hmC plays an important role in the second phase of DNA demethylation. Consistent with the TET1 and TET2 expression patterns, 5hmC level increases in E9.5–E10.5 PGCs with decreasing 5mC levels, peaks at E11.5, and then gradually declines at E11.5–E13.5 [88–91]. Genetic studies confirm the functions of TET1 in erasure of genomic imprinting and in activation of germline genes involved in gamete generation and meiosis [92–94]. Taken together, both passive and active demethylation pathways are involved in global demethylation, including the erasure of parental imprints, in PGCs. At E13.5, PGC genomes display the lowest level of global DNA methylation in the life cycle.

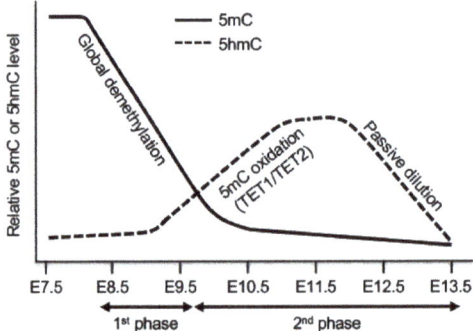

Figure 4. DNA demethylation in primordial germ cells (PGCs). PGCs undergo DNA demethylation in two phases. The first phase is mainly a passive process due to repression of important components of the DNA methylation machinery, resulting in global demethylation. The second phase, which affects specific loci including imprinted genes, is initiated by TET1- and TET2-mediated 5mC oxidation, followed by passive dilution of oxidized derivatives.

4.2. Sex-Specific Remethylation in the Germline

Sex-specific development of the mouse embryo starts at ~E12.5. Following erasure of DNA methylation marks in PGCs, male and female germ cells undergo remethylation at different time points during gametogenesis (Figure 1). While undergoing mitotic expansion in the developing gonad, the male germline starts the remethylation process as early as E14.5, and the methylation patterns are fully established by the time of birth. After sexual maturity, male germ cells undergo many rounds of mitotic proliferation to form spermatocytes before entering meiotic differentiation to form haploid spermatids. The sperm-specific methylation pattern is maintained during mitotic expansion [95]. The female germline undergoes mitotic expansion before primary oocytes arrest in prophase of meiosis I around the time of birth. After sexual maturity, oocytes complete meiosis I and arrest again in metaphase of meiosis II until fertilization. DNA methylation levels in primary oocytes before birth remain low, and remethylation occurs after birth in the oocyte growth phase [95]. In addition to differences in the timing of DNA methylation, the extent and distribution of DNA methylation are widely different between male and female germ cells. Sperm genomes are almost fully methylated (~90% of CpGs) except CGIs, whereas oocyte genomes show lower methylation levels (~40% of CpGs), with methylation marks being largely confined to intragenic regions of active genes [96].

4.3. Establishment of Methylation Imprints

Unlike most genes that show biallelic expression, imprinted genes are expressed from only one allele according to the parent of origin. Nearly 200 imprinted genes have been identified in mouse and 165 in human [97]. These genes are involved in many important processes, including embryogenesis, placenta formation, fetal and postnatal growth, and animal behavior [97,98]. Defects in imprinted gene expression are associated with reproductive disorders (e.g., infertility, hydatidiform mole) and congenital diseases (e.g., Prader–Willi syndrome, Angelman syndrome, Beckwith–Wiedemann syndrome, Silver–Russell syndrome) [99]. Loss of imprinting (i.e., biallelic expression or silencing of imprinted genes) is also frequently observed in cancer cells [100].

The majority of imprinted genes exist in clusters. In general, each cluster contains an ICR with one or more differentially methylated regions (DMRs) on the maternal and paternal alleles. These allele-specific methylation imprints are established in germ cells and, after fertilization, they are maintained during embryonic development and in somatic tissues.

As mentioned above, imprinted DMRs are erased in PGCs and subsequently remethylated, along with other genomic sequences, in a sex-specific manner. Most imprints are maternally derived, and their establishment occurs during the oocyte growth period after birth and is largely completed by the germinal vesicle stage of development and resumption of meiosis [66,101,102]. Only four paternally imprinted genes (i.e., *Igf2–H19*, *Dlk1–Gtl2*, *Rasgrf1*, and *Zdbf2*) are known. Paternal methylation imprints are acquired before birth and maintained through many cycles of mitotic division before entry into meiosis [99].

Genetic studies in mice demonstrate that DNMT3A is responsible for de novo methylation in the germline and its accessory factor DNMT3L is also required [56,64,103]. DNMT3A2 is the major isoform in developing germ cells [48,49]. DNMT3L may facilitate DNMT3A2 function by regulating DNMT3A2 enzymatic activity, stability, and targeting to specific genomic regions [59–63]. A proper chromatin environment is important for the DNMT3A2-DNMT3L complex to be recruited to its genomic targets, including imprinted loci. Biochemical and structural evidence indicates that H3K4 methylation disrupts the interaction between the DNMT3L/DNMT3A ADD domains and the N-terminal tail of histone H3 [45,63], which is supported by in vivo data obtained from genetic studies in mice. Oocytes deficient for KDM1B (also known as LSD2 and AOF1), an H3K4me1/2 demethylase, fail to establish imprinted DMRs at some maternally imprinted loci, as well as methylation marks in other regions, leading to a maternal-effect lethal phenotype [104,105]. A single-amino-acid mutation in the DNMT3L ADD domain causes overall reductions in DNA methylation, coupled with spermatogenesis defects [106]. Also, transcription across DMRs may create an open chromatin environment that facilitates imprint establishment in the germline. Truncation of transcript from an upstream promoter results in the absence of maternal methylation of DMRs at the *Gnas* locus in mouse oocytes [107].

5. DNA Methylation Reprogramming in Early Embryos

The purpose of epigenetic reprogramming during early embryogenesis is to acquire totipotency and create an epigenome for embryonic development. Key reprogramming events include the erasure of most DNA methylation marks inherited from the gametes in preimplantation embryos and the establishment of embryonic methylation patterns upon implantation (Figure 5).

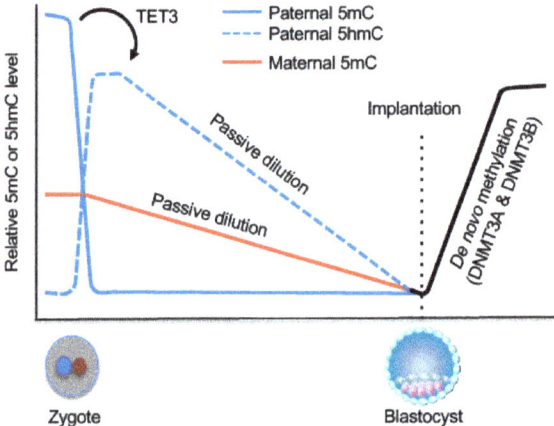

Figure 5. DNA demethylation and remethylation during early embryogenesis. Shortly after fertilization, TET3-mediated 5mC oxidation occurs in the paternal genome and the oxidized derivatives are subsequently removed through passive dilution during preimplantation development. The maternal genome mainly undergoes passive demethylation during preimplantation development. Upon implantation, a wave of de novo methylation establishes the embryonic methylation pattern.

5.1. DNA Demethylation During Preimplantation Development

In the zygote, the parental genomes are spatially separated in maternal and paternal pronuclei, with the maternal genome having a lower level of global DNA methylation than the paternal genome. The parental genomes then undergo genome-wide DNA demethylation with distinct mechanisms. As a consequence, blastocyst-stage embryos reach the lowest methylation levels. However, some regions escape this wave of demethylation, including some retrotransposons and the DMRs of imprinted loci (Figure 5).

Mature sperm genomes have the highest DNA methylation level of all cell types in mammals, with the vast majority (~90%) of CpGs being methylated. Earlier immunofluorescence data showed that the paternal genome loses the 5mC signal before the onset of DNA replication [108,109], suggesting that the paternal genome undergoes active demethylation. Subsequent work demonstrates that 5mC in the paternal genome is converted to 5hmC, 5fC, and 5caC by TET3, which is highly expressed in oocytes and thus abundant in zygotes [110,111]. Rather than being rapidly replaced by unmodified cytosine, 5hmC, 5fC, and 5caC persist in the paternal genome and undergo gradual decline during cleavage divisions [112,113], indicating that, although these 5mC derivatives are produced in the zygote by an enzyme-catalyzed process, their loss during preimplantation development is mainly through a DNA replication-coupled passive process.

The maternal and paternal genomes are exposed to an identical environment in the zygote, yet the maternal genome is protected from TET3-mediated 5mC oxidation. STELLA (also known as DPPA3 and PGC7), a maternal factor, is required for this protection by inhibiting the recruitment of TET3 to the maternal pronucleus [114]. Loss of 5mC in the maternal genome mainly results from DNA replication-dependent dilution during preimplantation development, presumably due to DNMT1 exclusion from the nuclei [115,116]. In preimplantation embryos, DNMT1o, transmitted from oocytes, is the predominant DNMT1 isoform [117], which lacks the N-terminal 118 amino acids of full-length DNMT1. However, the absence of this region does not cause the localization change, as ectopically expressed DNMT1o in somatic cells localizes in the nuclei [115]. The mechanism underlying DNMT1o cytoplasmic localization in early embryos remains to be determined.

5.2. Maintenance of Methylation Imprints

Despite the extensive DNA demethylation during preimplantation development, methylation marks in DMRs in imprinted loci are spared. Genetic evidence reveals the requirement of *Dnmt1*, but not *Dnmt3a* and *Dnmt3b*, in the maintenance of methylation imprints [117]. Thus, even though the majority of DNMT1 is excluded from the nucleus, it is not exclusively retained in the cytoplasm. While how the limited amount of DNMT1 is targeted to imprinted loci is not fully understood, the ZFP57-TRIM28 complex plays a critical role. ZFP57 is a Kruppel-like zinc finger protein that binds a consensus hexanucleotide (TGCCGC) present in ICRs, but only when the CpG site is methylated [118]. Removal of ZFP57 from the mouse zygote results in embryonic and neonatal lethality due to defects in the maintenance of both maternal and paternal imprints at multiple imprinted loci [119]. The function of ZFP57 is evolutionarily conserved, as loss-of-function mutations in human are associated with ICR hypomethylation at multiple imprinted loci, resulting in transient neonatal diabetes [120]. TRIM28 (also known as KAP1) is a scaffold protein that assembles a heterochromatin-inducing complex, comprised of SETDB1 (a histone methyltransferase catalyzing H3K9me3), the nucleosome remodeling and histone deacetylation (NuRD) complex, heterochromatin protein 1 (HP1), and the DNA methylation machinery including DNMT1 and UHRF1 [121]. Deletion of *Trim28* in the maternal germ line leads to hypomethylation at several maternal and paternal ICRs and embryonic lethality [121]. In addition to DNMT1-mediated maintenance of methylation marks in ICRs, mechanisms that prevent them from being erased are also important. STELLA has been shown to protect methylation of maternally imprinted genes, as well as two paternally imprinted genes, *H19* and *Rasgrf1* [122], presumably by binding to H3K9me2 to block the recruitment of TET3. During spermatogenesis, the majority of the core histones are replaced by protamines. There is evidence that H3K9me2 is present at the DMRs of *H19* and *Rasgrf1* during protamine exchange [114].

5.3. De novo Methylation in the Epiblast

After implantation, a wave of de novo methylation occurs in the epiblast to establish the new methylation pattern in the embryo and start a new life cycle. Genome-wide analysis of mouse embryos from blastocyst to post-implantation stages reveals that DNA methylation is established within two days of implantation between E4.5 and E6.5 [123]. De novo methylation is catalyzed by DNMT3A and DNMT3B. *Dnmt3a* KO mice survive to term but become runted and die within four weeks after birth, inactivation of *Dnmt3b* results in embryonic lethality after E12.5, and embryos deficient for both *Dnmt3a* and *Dnmt3b* exhibit a more severe developmental phenotype and die earlier than *Dnmt3b* KO embryos [50]. In agreement with the developmental phenotypes, DNA methylation analysis of mutant embryos indicates that DNMT3A and DNMT3B act redundantly in methylating the bulk genome and repetitive elements, whereas DNMT3B has a prominent role in methylating CpG islands [123,124]. While what specify DNA methylation patterns is not fully understood, chromatin structure is likely a major determinant. Numerous histone modifications and enzymes/regulators involved in chromatin structure positively or negatively modulate DNA methylation [125,126]. Despite the essential role of DNMT3L in DNA methylation in the germline, it is not required for embryonic development [56,64,65]. There is indication that *Dnmt3l* deficiency causes lower levels of DNA methylation in early embryos, which are subsequently recovered during development [62,127].

6. Future Directions

The purpose of the two waves of epigenetic reprogramming is to create appropriate epigenetic states for the initiation and development of the next generation. Among the epigenetic reprogramming events, dynamic changes in DNA methylation, including global erasure and subsequent re-establishment, during germ cell development and early embryogenesis are well documented. Over the last three decades, great progress has been made in understanding the DNA methylation and demethylation machineries, including the identification and characterization of DNMTs, TETs, and key regulators. However, numerous important questions regarding DNA methylation reprogramming remain to be answered. For example, passive demethylation of the maternal genome during preimplantation development is presumably due to exclusion of DNMT1 from the nuclei. Little is known about the mechanisms by which DNMT1 is excluded from the nuclei and/or sequestered in the cytoplasm. Studies in the last several years indicate that demethylation of the paternal genome is initiated by TET3-mediated 5mC oxidation, which is considered an 'active' process, but the loss of 5hmC, 5fC, and 5caC is mainly achieved via DNA replication-dependent 'passive' dilution. It remains to be determined why these 5mC oxidized derivatives can be maintained at steady-state levels in ESCs and somatic cells, but not in cells of preimplantation embryos. Another issue that is poorly understood is the determinants of sex-specific DNA methylation during germ cell development, including the establishment of methylation imprints. Specifically, it is largely unknown why male and female germ cells are differentially methylated, despite their almost identical genomes (except sex chromosomes). Comprehensive understanding of epigenetic reprogramming in vivo will facilitate development of experimental reprogramming in vitro and could also be valuable for developing novel therapeutic strategies for infertility, imprinting diseases, and other developmental disorders.

Author Contributions: Y.Z. and T.C. wrote the manuscript.

Funding: Work in our laboratory is supported by grants from the National Institutes of Health (1R01DK106418-01 and 1R01AI12140301A1 to T.C.). Y.Z. is a recipient of the Thomas Endowment fellowship.

Conflicts of Interest: The authors declare no conflict of interest.

References

1. Patil, V.; Ward, R.L.; Hesson, L.B. The evidence for functional non-CpG methylation in mammalian cells. *Epigenetics* **2014**, *9*, 823–828. [CrossRef] [PubMed]
2. Deaton, A.M.; Bird, A. CpG islands and the regulation of transcription. *Genes Dev.* **2011**, *25*, 1010–1022. [CrossRef] [PubMed]
3. Smith, Z.D.; Meissner, A. DNA methylation: Roles in mammalian development. *Nat. Rev. Genet.* **2013**, *14*, 204–220. [CrossRef] [PubMed]
4. Hamidi, T.; Singh, A.K.; Chen, T. Genetic alterations of DNA methylation machinery in human diseases. *Epigenomics* **2015**, *7*, 247–265. [CrossRef]
5. Baylin, S.B.; Jones, P.A. Epigenetic Determinants of Cancer. *Cold. Spring Harb. Perspect. Biol.* **2016**, *8*, a019505. [CrossRef]
6. Smallwood, S.A.; Kelsey, G. De novo DNA methylation: a germ cell perspective. *Trends Genet* **2012**, *28*, 33–42. [CrossRef]
7. Messerschmidt, D.M.; Knowles, B.B.; Solter, D. DNA methylation dynamics during epigenetic reprogramming in the germline and preimplantation embryos. *Genes Dev.* **2014**, *28*, 812–828. [CrossRef]
8. Holliday, R.; Pugh, J.E. DNA modification mechanisms and gene activity during development. *Science* **1975**, *187*, 226–232. [CrossRef]
9. Riggs, A.D. X inactivation, differentiation, and DNA methylation. *Cytogenet. Cell Genet.* **1975**, *14*, 9–25. [CrossRef]
10. Chen, T.; Li, E. Structure and function of eukaryotic DNA methyltransferases. *Curr. Top. Dev. Biol.* **2004**, *60*, 55–89. [PubMed]
11. Bestor, T.; Laudano, A.; Mattaliano, R.; Ingram, V. Cloning and sequencing of a cDNA encoding DNA methyltransferase of mouse cells. The carboxyl-terminal domain of the mammalian enzymes is related to bacterial restriction methyltransferases. *J. Mol. Biol.* **1988**, *203*, 971–983. [CrossRef]
12. Mertineit, C.; Yoder, J.A.; Taketo, T.; Laird, D.W.; Trasler, J.M.; Bestor, T.H. Sex-specific exons control DNA methyltransferase in mammalian germ cells. *Development* **1998**, *125*, 889–897.
13. Rouleau, J.; Tanigawa, G.; Szyf, M. The mouse DNA methyltransferase 5'-region. A unique housekeeping gene promoter. *J. Biol. Chem.* **1992**, *267*, 7368–7377.

14. Ding, F.; Chaillet, J.R. In vivo stabilization of the Dnmt1 (cytosine-5)- methyltransferase protein. *Proc. Natl. Acad. Sci. USA* **2002**, *99*, 14861–14866. [CrossRef] [PubMed]
15. Takeshita, K.; Suetake, I.; Yamashita, E.; Suga, M.; Narita, H.; Nakagawa, A.; Tajima, S. Structural insight into maintenance methylation by mouse DNA methyltransferase 1 (Dnmt1). *Proc. Natl. Acad. Sci. USA* **2011**, *108*, 9055–9059. [CrossRef] [PubMed]
16. Pradhan, S.; Bacolla, A.; Wells, R.D.; Roberts, R.J. Recombinant human DNA (cytosine-5) methyltransferase. I. Expression, purification, and comparison of de novo and maintenance methylation. *J. Biol. Chem.* **1999**, *274*, 33002–33010. [CrossRef] [PubMed]
17. Kishikawa, S.; Murata, T.; Ugai, H.; Yamazaki, T.; Yokoyama, K.K. Control elements of *Dnmt1* gene are regulated in cell-cycle dependent manner. *Nucleic Acids Res. Suppl.* **2003**, *3*, 307–308. [CrossRef]
18. Leonhardt, H.; Page, A.W.; Weier, H.U.; Bestor, T.H. A targeting sequence directs DNA methyltransferase to sites of DNA replication in mammalian nuclei. *Cell* **1992**, *71*, 865–873. [CrossRef]
19. Li, E.; Bestor, T.H.; Jaenisch, R. Targeted mutation of the DNA methyltransferase gene results in embryonic lethality. *Cell* **1992**, *69*, 915–926. [CrossRef]
20. Lei, H.; Oh, S.P.; Okano, M.; Juttermann, R.; Goss, K.A.; Jaenisch, R.; Li, E. De novo DNA cytosine methyltransferase activities in mouse embryonic stem cells. *Development* **1996**, *122*, 3195–3205. [PubMed]
21. Li, Y.; Zhang, Z.; Chen, J.; Liu, W.; Lai, W.; Liu, B.; Li, X.; Liu, L.; Xu, S.; Dong, Q.; et al. Stella safeguards the oocyte methylome by preventing de novo methylation mediated by DNMT1. *Nature* **2018**, *564*, 136–140. [CrossRef]
22. Bostick, M.; Kim, J.K.; Esteve, P.O.; Clark, A.; Pradhan, S.; Jacobsen, S.E. UHRF1 plays a role in maintaining DNA methylation in mammalian cells. *Science* **2007**, *317*, 1760–1764. [CrossRef]
23. Sharif, J.; Muto, M.; Takebayashi, S.; Suetake, I.; Iwamatsu, A.; Endo, T.A.; Shinga, J.; Mizutani-Koseki, Y.; Toyoda, T.; Okamura, K.; et al. The SRA protein Np95 mediates epigenetic inheritance by recruiting Dnmt1 to methylated DNA. *Nature* **2007**, *450*, 908–912. [CrossRef] [PubMed]
24. Liu, X.; Gao, Q.; Li, P.; Zhao, Q.; Zhang, J.; Li, J.; Koseki, H.; Wong, J. UHRF1 targets DNMT1 for DNA methylation through cooperative binding of hemi-methylated DNA and methylated H3K9. *Nat. Commun.* **2013**, *4*, 1563. [CrossRef] [PubMed]
25. Rothbart, S.B.; Krajewski, K.; Nady, N.; Tempel, W.; Xue, S.; Badeaux, A.I.; Barsyte-Lovejoy, D.; Martinez, J.Y.; Bedford, M.T.; Fuchs, S.M.; et al. Association of UHRF1 with methylated H3K9 directs the maintenance of DNA methylation. *Nat. Struct. Mol. Biol.* **2012**, *19*, 1155–1160. [CrossRef] [PubMed]
26. Rothbart, S.B.; Dickson, B.M.; Ong, M.S.; Krajewski, K.; Houliston, S.; Kireev, D.B.; Arrowsmith, C.H.; Strahl, B.D. Multivalent histone engagement by the linked tandem Tudor and PHD domains of UHRF1 is required for the epigenetic inheritance of DNA methylation. *Genes Dev.* **2013**, *27*, 1288–1298. [CrossRef]
27. Rajakumara, E.; Wang, Z.; Ma, H.; Hu, L.; Chen, H.; Lin, Y.; Guo, R.; Wu, F.; Li, H.; Lan, F.; et al. PHD finger recognition of unmodified histone H3R2 links UHRF1 to regulation of euchromatic gene expression. *Mol. Cell* **2011**, *43*, 275–284. [CrossRef] [PubMed]
28. Hu, L.; Li, Z.; Wang, P.; Lin, Y.; Xu, Y. Crystal structure of PHD domain of UHRF1 and insights into recognition of unmodified histone H3 arginine residue 2. *Cell Res.* **2011**, *21*, 1374–1378. [CrossRef] [PubMed]
29. Wang, C.; Shen, J.; Yang, Z.; Chen, P.; Zhao, B.; Hu, W.; Lan, W.; Tong, X.; Wu, H.; Li, G.; et al. Structural basis for site-specific reading of unmodified R2 of histone H3 tail by UHRF1 PHD finger. *Cell Res.* **2011**, *21*, 1379–1382. [CrossRef]
30. Nishiyama, A.; Yamaguchi, L.; Sharif, J.; Johmura, Y.; Kawamura, T.; Nakanishi, K.; Shimamura, S.; Arita, K.; Kodama, T.; Ishikawa, F.; et al. Uhrf1-dependent H3K23 ubiquitylation couples maintenance DNA methylation and replication. *Nature* **2013**, *502*, 249–253. [CrossRef]
31. Qin, W.; Wolf, P.; Liu, N.; Link, S.; Smets, M.; La Mastra, F.; Forne, I.; Pichler, G.; Horl, D.; Fellinger, K.; et al. DNA methylation requires a DNMT1 ubiquitin interacting motif (UIM) and histone ubiquitination. *Cell Res.* **2015**, *25*, 911–929. [CrossRef] [PubMed]
32. Harrison, J.S.; Cornett, E.M.; Goldfarb, D.; DaRosa, P.A.; Li, Z.M.; Yan, F.; Dickson, B.M.; Guo, A.H.; Cantu, D.V.; Kaustov, L.; et al. Hemi-methylated DNA regulates DNA methylation inheritance through allosteric activation of H3 ubiquitylation by UHRF1. *Elife* **2016**, *5*, e17101. [CrossRef]
33. Fang, J.; Cheng, J.; Wang, J.; Zhang, Q.; Liu, M.; Gong, R.; Wang, P.; Zhang, X.; Feng, Y.; Lan, W.; et al. Hemi-methylated DNA opens a closed conformation of UHRF1 to facilitate its histone recognition. *Nat. Commun.* **2016**, *7*, 11197. [CrossRef]

34. Foster, B.M.; Stolz, P.; Mulholland, C.B.; Montoya, A.; Kramer, H.; Bultmann, S.; Bartke, T. Critical Role of the UBL Domain in Stimulating the E3 Ubiquitin Ligase Activity of UHRF1 toward Chromatin. *Mol. Cell* **2018**, *72*, 739–752. [CrossRef] [PubMed]
35. DaRosa, P.A.; Harrison, J.S.; Zelter, A.; Davis, T.N.; Brzovic, P.; Kuhlman, B.; Klevit, R.E. A Bifunctional Role for the UHRF1 UBL Domain in the Control of Hemi-methylated DNA-Dependent Histone Ubiquitylation. *Mol. Cell* **2018**, *72*, 753–765. [CrossRef]
36. Okano, M.; Xie, S.; Li, E. Cloning and characterization of a family of novel mammalian DNA (cytosine-5) methyltransferases. *Nat. Genet.* **1998**, *19*, 219–220. [CrossRef] [PubMed]
37. Chen, T.; Ueda, Y.; Xie, S.; Li, E. A novel Dnmt3a isoform produced from an alternative promoter localizes to euchromatin and its expression correlates with active de novo methylation. *J. Biol. Chem.* **2002**, *277*, 38746–38754. [CrossRef] [PubMed]
38. Ostler, K.R.; Yang, Q.; Looney, T.J.; Zhang, L.; Vasanthakumar, A.; Tian, Y.; Kocherginsky, M.; Raimondi, S.L.; DeMaio, J.G.; Salwen, H.R.; et al. Truncated DNMT3B isoform DNMT3B7 suppresses growth, induces differentiation, and alters DNA methylation in human neuroblastoma. *Cancer Res.* **2012**, *72*, 4714–4723. [CrossRef]
39. Duymich, C.E.; Charlet, J.; Yang, X.; Jones, P.A.; Liang, G. DNMT3B isoforms without catalytic activity stimulate gene body methylation as accessory proteins in somatic cells. *Nat. Commun.* **2016**, *7*, 11453. [CrossRef]
40. Qiu, C.; Sawada, K.; Zhang, X.; Cheng, X. The PWWP domain of mammalian DNA methyltransferase Dnmt3b defines a new family of DNA-binding folds. *Nat. Struct. Biol.* **2002**, *9*, 217–224. [CrossRef] [PubMed]
41. Chen, T.; Tsujimoto, N.; Li, E. The PWWP domain of Dnmt3a and Dnmt3b is required for directing DNA methylation to the major satellite repeats at pericentric heterochromatin. *Mol. Cell. Biol.* **2004**, *24*, 9048–9058. [CrossRef]
42. Dhayalan, A.; Rajavelu, A.; Rathert, P.; Tamas, R.; Jurkowska, R.Z.; Ragozin, S.; Jeltsch, A. The Dnmt3a PWWP domain reads histone 3 lysine 36 trimethylation and guides DNA methylation. *J. Biol. Chem.* **2010**, *285*, 26114–26120. [CrossRef] [PubMed]
43. Baubec, T.; Colombo, D.F.; Wirbelauer, C.; Schmidt, J.; Burger, L.; Krebs, A.R.; Akalin, A.; Schubeler, D. Genomic profiling of DNA methyltransferases reveals a role for DNMT3B in genic methylation. *Nature* **2015**, *520*, 243–247. [CrossRef]
44. Rondelet, G.; Dal Maso, T.; Willems, L.; Wouters, J. Structural basis for recognition of histone H3K36me3 nucleosome by human de novo DNA methyltransferases 3A and 3B. *J. Struct. Biol.* **2016**, *194*, 357–367. [CrossRef] [PubMed]
45. Otani, J.; Nankumo, T.; Arita, K.; Inamoto, S.; Ariyoshi, M.; Shirakawa, M. Structural basis for recognition of H3K4 methylation status by the DNA methyltransferase 3A ATRX-DNMT3-DNMT3L domain. *EMBO Rep.* **2009**, *10*, 1235–1241. [CrossRef]
46. Guo, X.; Wang, L.; Li, J.; Ding, Z.; Xiao, J.; Yin, X.; He, S.; Shi, P.; Dong, L.; Li, G.; et al. Structural insight into autoinhibition and histone H3-induced activation of DNMT3A. *Nature* **2015**, *517*, 640–644. [CrossRef]
47. Watanabe, D.; Suetake, I.; Tada, T.; Tajima, S. Stage- and cell-specific expression of Dnmt3a and Dnmt3b during embryogenesis. *Mech. Dev.* **2002**, *118*, 187–190. [CrossRef]
48. Sakai, Y.; Suetake, I.; Shinozaki, F.; Yamashina, S.; Tajima, S. Co-expression of de novo DNA methyltransferases Dnmt3a2 and Dnmt3L in gonocytes of mouse embryos. *Gene Expr. Patterns* **2004**, *5*, 231–237. [CrossRef]
49. Ma, P.; de Waal, E.; Weaver, J.R.; Bartolomei, M.S.; Schultz, R.M. A DNMT3A2-HDAC2 Complex Is Essential for Genomic Imprinting and Genome Integrity in Mouse Oocytes. *Cell Rep.* **2015**, *13*, 1552–1560. [CrossRef]
50. Okano, M.; Bell, D.W.; Haber, D.A.; Li, E. DNA methyltransferases Dnmt3a and Dnmt3b are essential for de novo methylation and mammalian development. *Cell* **1999**, *99*, 247–257. [CrossRef]
51. Chen, T.; Ueda, Y.; Dodge, J.E.; Wang, Z.; Li, E. Establishment and maintenance of genomic methylation patterns in mouse embryonic stem cells by Dnmt3a and Dnmt3b. *Mol. Cell. Biol.* **2003**, *23*, 5594–5605. [CrossRef] [PubMed]
52. Lees-Murdock, D.J.; McLoughlin, G.A.; McDaid, J.R.; Quinn, L.M.; O'Doherty, A.; Hiripi, L.; Hack, C.J.; Walsh, C.P. Identification of 11 pseudogenes in the DNA methyltransferase gene family in rodents and humans with implications for the functional loci. *Genomics* **2004**, *84*, 193–204. [CrossRef] [PubMed]

53. Barau, J.; Teissandier, A.; Zamudio, N.; Roy, S.; Nalesso, V.; Herault, Y.; Guillou, F.; Bourc'his, D. The DNA methyltransferase DNMT3C protects male germ cells from transposon activity. *Science* **2016**, *354*, 909–912. [CrossRef] [PubMed]
54. Aapola, U.; Kawasaki, K.; Scott, H.S.; Ollila, J.; Vihinen, M.; Heino, M.; Shintani, A.; Minoshima, S.; Krohn, K.; Antonarakis, S.E.; et al. Isolation and initial characterization of a novel zinc finger gene, *DNMT3L*, on 21q22.3, related to the cytosine-5-methyltransferase 3 gene family. *Genomics* **2000**, *65*, 293–298. [CrossRef] [PubMed]
55. Aapola, U.; Lyle, R.; Krohn, K.; Antonarakis, S.E.; Peterson, P. Isolation and initial characterization of the mouse *Dnmt3l* gene. *Cytogenet. Cell Genet.* **2001**, *92*, 122–126. [CrossRef] [PubMed]
56. Hata, K.; Okano, M.; Lei, H.; Li, E. Dnmt3L cooperates with the Dnmt3 family of de novo DNA methyltransferases to establish maternal imprints in mice. *Development* **2002**, *129*, 1983–1993. [PubMed]
57. Jia, D.; Jurkowska, R.Z.; Zhang, X.; Jeltsch, A.; Cheng, X. Structure of Dnmt3a bound to Dnmt3L suggests a model for de novo DNA methylation. *Nature* **2007**, *449*, 248–251. [CrossRef] [PubMed]
58. Zhang, Z.M.; Lu, R.; Wang, P.; Yu, Y.; Chen, D.; Gao, L.; Liu, S.; Ji, D.; Rothbart, S.B.; Wang, Y.; et al. Structural basis for DNMT3A-mediated de novo DNA methylation. *Nature* **2018**, *554*, 387–391. [CrossRef] [PubMed]
59. Chedin, F.; Lieber, M.R.; Hsieh, C.L. The DNA methyltransferase-like protein DNMT3L stimulates de novo methylation by Dnmt3a. *Proc. Natl. Acad. Sci. USA* **2002**, *99*, 16916–16921. [CrossRef]
60. Suetake, I.; Shinozaki, F.; Miyagawa, J.; Takeshima, H.; Tajima, S. DNMT3L stimulates the DNA methylation activity of Dnmt3a and Dnmt3b through a direct interaction. *J. Biol. Chem.* **2004**, *279*, 27816–27823. [CrossRef]
61. Gowher, H.; Liebert, K.; Hermann, A.; Xu, G.; Jeltsch, A. Mechanism of stimulation of catalytic activity of Dnmt3A and Dnmt3B DNA-(cytosine-C5)-methyltransferases by Dnmt3L. *J. Biol. Chem.* **2005**, *280*, 13341–13348. [CrossRef] [PubMed]
62. Veland, N.; Lu, Y.; Hardikar, S.; Gaddis, S.; Zeng, Y.; Liu, B.; Estecio, M.R.; Takata, Y.; Lin, K.; Tomida, M.W.; et al. DNMT3L facilitates DNA methylation partly by maintaining DNMT3A stability in mouse embryonic stem cells. *Nucleic Acids Res.* **2019**, *47*, 152–167. [CrossRef] [PubMed]
63. Ooi, S.K.; Qiu, C.; Bernstein, E.; Li, K.; Jia, D.; Yang, Z.; Erdjument-Bromage, H.; Tempst, P.; Lin, S.P.; Allis, C.D.; et al. DNMT3L connects unmethylated lysine 4 of histone H3 to de novo methylation of DNA. *Nature* **2007**, *448*, 714–717. [CrossRef] [PubMed]
64. Bourc'his, D.; Xu, G.L.; Lin, C.S.; Bollman, B.; Bestor, T.H. Dnmt3L and the establishment of maternal genomic imprints. *Science* **2001**, *294*, 2536–2539. [CrossRef]
65. Bourc'his, D.; Bestor, T.H. Meiotic catastrophe and retrotransposon reactivation in male germ cells lacking Dnmt3L. *Nature* **2004**, *431*, 96–99. [CrossRef]
66. Smallwood, S.A.; Tomizawa, S.; Krueger, F.; Ruf, N.; Carli, N.; Segonds-Pichon, A.; Sato, S.; Hata, K.; Andrews, S.R.; Kelsey, G. Dynamic CpG island methylation landscape in oocytes and preimplantation embryos. *Nat. Genet.* **2011**, *43*, 811–814. [CrossRef]
67. Shirane, K.; Toh, H.; Kobayashi, H.; Miura, F.; Chiba, H.; Ito, T.; Kono, T.; Sasaki, H. Mouse oocyte methylomes at base resolution reveal genome-wide accumulation of non-CpG methylation and role of DNA methyltransferases. *PLoS Genet.* **2013**, *9*, e1003439. [CrossRef] [PubMed]
68. Neri, F.; Krepelova, A.; Incarnato, D.; Maldotti, M.; Parlato, C.; Galvagni, F.; Matarese, F.; Stunnenberg, H.G.; Oliviero, S. Dnmt3L antagonizes DNA methylation at bivalent promoters and favors DNA methylation at gene bodies in ESCs. *Cell* **2013**, *155*, 121–134. [CrossRef]
69. Okano, M.; Xie, S.; Li, E. Dnmt2 is not required for de novo and maintenance methylation of viral DNA in embryonic stem cells. *Nucleic Acids Res.* **1998**, *26*, 2536–2540. [CrossRef] [PubMed]
70. Goll, M.G.; Kirpekar, F.; Maggert, K.A.; Yoder, J.A.; Hsieh, C.L.; Zhang, X.; Golic, K.G.; Jacobsen, S.E.; Bestor, T.H. Methylation of tRNAAsp by the DNA methyltransferase homolog Dnmt2. *Science* **2006**, *311*, 395–398. [CrossRef]
71. Zhang, Y.; Zhang, X.; Shi, J.; Tuorto, F.; Li, X.; Liu, Y.; Liebers, R.; Zhang, L.; Qu, Y.; Qian, J.; et al. Dnmt2 mediates intergenerational transmission of paternally acquired metabolic disorders through sperm small non-coding RNAs. *Nat. Cell Biol.* **2018**, *20*, 535–540. [CrossRef]
72. Wu, H.; Zhang, Y. Reversing DNA methylation: Mechanisms, genomics, and biological functions. *Cell* **2014**, *156*, 45–68. [CrossRef]
73. Kagiwada, S.; Kurimoto, K.; Hirota, T.; Yamaji, M.; Saitou, M. Replication-coupled passive DNA demethylation for the erasure of genome imprints in mice. *EMBO J.* **2013**, *32*, 340–353. [CrossRef]

74. Dan, J.; Rousseau, P.; Hardikar, S.; Veland, N.; Wong, J.; Autexier, C.; Chen, T. Zscan4 Inhibits Maintenance DNA Methylation to Facilitate Telomere Elongation in Mouse Embryonic Stem Cells. *Cell Rep.* **2017**, *20*, 1936–1949. [CrossRef] [PubMed]
75. Veland, N.; Hardikar, S.; Zhong, Y.; Gayatri, S.; Dan, J.; Strahl, B.D.; Rothbart, S.B.; Bedford, M.T.; Chen, T. The Arginine Methyltransferase PRMT6 Regulates DNA Methylation and Contributes to Global DNA Hypomethylation in Cancer. *Cell Rep.* **2017**, *21*, 3390–3397. [CrossRef] [PubMed]
76. Tahiliani, M.; Koh, K.P.; Shen, Y.; Pastor, W.A.; Bandukwala, H.; Brudno, Y.; Agarwal, S.; Iyer, L.M.; Liu, D.R.; Aravind, L.; et al. Conversion of 5-methylcytosine to 5-hydroxymethylcytosine in mammalian DNA by MLL partner TET1. *Science* **2009**, *324*, 930–935. [CrossRef] [PubMed]
77. Ito, S.; D'Alessio, A.C.; Taranova, O.V.; Hong, K.; Sowers, L.C.; Zhang, Y. Role of Tet proteins in 5mC to 5hmC conversion, ES-cell self-renewal and inner cell mass specification. *Nature* **2010**, *466*, 1129–1133. [CrossRef] [PubMed]
78. Ito, S.; Shen, L.; Dai, Q.; Wu, S.C.; Collins, L.B.; Swenberg, J.A.; He, C.; Zhang, Y. Tet proteins can convert 5-methylcytosine to 5-formylcytosine and 5-carboxylcytosine. *Science* **2011**, *333*, 1300–1303. [CrossRef] [PubMed]
79. He, Y.F.; Li, B.Z.; Li, Z.; Liu, P.; Wang, Y.; Tang, Q.; Ding, J.; Jia, Y.; Chen, Z.; Li, L.; et al. Tet-mediated formation of 5-carboxylcytosine and its excision by TDG in mammalian DNA. *Science* **2011**, *333*, 1303–1307. [CrossRef] [PubMed]
80. Hu, L.; Lu, J.; Cheng, J.; Rao, Q.; Li, Z.; Hou, H.; Lou, Z.; Zhang, L.; Li, W.; Gong, W.; et al. Structural insight into substrate preference for TET-mediated oxidation. *Nature* **2015**, *527*, 118–122. [CrossRef]
81. Ko, M.; An, J.; Bandukwala, H.S.; Chavez, L.; Aijo, T.; Pastor, W.A.; Segal, M.F.; Li, H.; Koh, K.P.; Lahdesmaki, H.; et al. Modulation of TET2 expression and 5-methylcytosine oxidation by the CXXC domain protein IDAX. *Nature* **2013**. [CrossRef]
82. Maiti, A.; Drohat, A.C. Thymine DNA glycosylase can rapidly excise 5-formylcytosine and 5-carboxylcytosine: Potential implications for active demethylation of CpG sites. *J. Biol. Chem.* **2011**, *286*, 35334–35338. [CrossRef] [PubMed]
83. Zhang, L.; Lu, X.; Lu, J.; Liang, H.; Dai, Q.; Xu, G.L.; Luo, C.; Jiang, H.; He, C. Thymine DNA glycosylase specifically recognizes 5-carboxylcytosine-modified DNA. *Nat. Chem. Biol.* **2012**, *8*, 328–330. [CrossRef] [PubMed]
84. Hashimoto, H.; Liu, Y.; Upadhyay, A.K.; Chang, Y.; Howerton, S.B.; Vertino, P.M.; Zhang, X.; Cheng, X. Recognition and potential mechanisms for replication and erasure of cytosine hydroxymethylation. *Nucleic Acids Res.* **2012**, *40*, 4841–4849. [CrossRef]
85. Saitou, M. Germ cell specification in mice. *Curr. Opin. Genet. Dev.* **2009**, *19*, 386–395. [CrossRef] [PubMed]
86. Guibert, S.; Forne, T.; Weber, M. Global profiling of DNA methylation erasure in mouse primordial germ cells. *Genome Res.* **2012**, *22*, 633–641. [CrossRef] [PubMed]
87. Seisenberger, S.; Andrews, S.; Krueger, F.; Arand, J.; Walter, J.; Santos, F.; Popp, C.; Thienpont, B.; Dean, W.; Reik, W. The dynamics of genome-wide DNA methylation reprogramming in mouse primordial germ cells. *Mol. Cell* **2012**, *48*, 849–862. [CrossRef] [PubMed]
88. Hackett, J.A.; Sengupta, R.; Zylicz, J.J.; Murakami, K.; Lee, C.; Down, T.A.; Surani, M.A. Germline DNA demethylation dynamics and imprint erasure through 5-hydroxymethylcytosine. *Science* **2013**, *339*, 448–452. [CrossRef]
89. Yamaguchi, S.; Hong, K.; Liu, R.; Inoue, A.; Shen, L.; Zhang, K.; Zhang, Y. Dynamics of 5-methylcytosine and 5-hydroxymethylcytosine during germ cell reprogramming. *Cell Res.* **2013**, *23*, 329–339. [CrossRef]
90. Vincent, J.J.; Huang, Y.; Chen, P.Y.; Feng, S.; Calvopina, J.H.; Nee, K.; Lee, S.A.; Le, T.; Yoon, A.J.; Faull, K.; et al. Stage-specific roles for Tet1 and Tet2 in DNA demethylation in primordial germ cells. *Cell Stem Cell* **2013**, *12*, 470–478. [CrossRef]
91. Piccolo, F.M.; Bagci, H.; Brown, K.E.; Landeira, D.; Soza-Ried, J.; Feytout, A.; Mooijman, D.; Hajkova, P.; Leitch, H.G.; Tada, T.; et al. Different Roles for Tet1 and Tet2 Proteins in Reprogramming-Mediated Erasure of Imprints Induced by EGC Fusion. *Mol. Cell* **2013**, *49*, 1023–1033. [CrossRef] [PubMed]
92. Yamaguchi, S.; Hong, K.; Liu, R.; Shen, L.; Inoue, A.; Diep, D.; Zhang, K.; Zhang, Y. Tet1 controls meiosis by regulating meiotic gene expression. *Nature* **2012**, *492*, 443–447. [CrossRef]
93. Yamaguchi, S.; Shen, L.; Liu, Y.; Sendler, D.; Zhang, Y. Role of Tet1 in erasure of genomic imprinting. *Nature* **2013**, *504*, 460–464. [CrossRef]

94. Hill, P.W.S.; Leitch, H.G.; Requena, C.E.; Sun, Z.; Amouroux, R.; Roman-Trufero, M.; Borkowska, M.; Terragni, J.; Vaisvila, R.; Linnett, S.; et al. Epigenetic reprogramming enables the transition from primordial germ cell to gonocyte. *Nature* **2018**, *555*, 392–396. [CrossRef] [PubMed]
95. Kota, S.K.; Feil, R. Epigenetic transitions in germ cell development and meiosis. *Dev. Cell* **2010**, *19*, 675–686. [CrossRef]
96. Kobayashi, H.; Sakurai, T.; Imai, M.; Takahashi, N.; Fukuda, A.; Yayoi, O.; Sato, S.; Nakabayashi, K.; Hata, K.; Sotomaru, Y.; et al. Contribution of intragenic DNA methylation in mouse gametic DNA methylomes to establish oocyte-specific heritable marks. *PLoS Genet.* **2012**, *8*, e1002440. [CrossRef]
97. Tucci, V.; Isles, A.R.; Kelsey, G.; Ferguson-Smith, A.C.; Erice Imprinting, G. Genomic Imprinting and Physiological Processes in Mammals. *Cell* **2019**, *176*, 952–965. [CrossRef] [PubMed]
98. Reik, W.; Walter, J. Genomic imprinting: Parental influence on the genome. *Nat. Rev. Genet.* **2001**, *2*, 21–32. [CrossRef] [PubMed]
99. Monk, D.; Mackay, D.J.G.; Eggermann, T.; Maher, E.R.; Riccio, A. Genomic imprinting disorders: Lessons on how genome, epigenome and environment interact. *Nat. Rev. Genet.* **2019**. [CrossRef]
100. Goovaerts, T.; Steyaert, S.; Vandenbussche, C.A.; Galle, J.; Thas, O.; Van Criekinge, W.; De Meyer, T. A comprehensive overview of genomic imprinting in breast and its deregulation in cancer. *Nat. Commun.* **2018**, *9*, 4120. [CrossRef]
101. Hiura, H.; Obata, Y.; Komiyama, J.; Shirai, M.; Kono, T. Oocyte growth-dependent progression of maternal imprinting in mice. *Genes Cells* **2006**, *11*, 353–361. [CrossRef] [PubMed]
102. Gahurova, L.; Tomizawa, S.I.; Smallwood, S.A.; Stewart-Morgan, K.R.; Saadeh, H.; Kim, J.; Andrews, S.R.; Chen, T.; Kelsey, G. Transcription and chromatin determinants of de novo DNA methylation timing in oocytes. *Epigenet. Chromatin* **2017**, *10*, 25. [CrossRef]
103. Kaneda, M.; Okano, M.; Hata, K.; Sado, T.; Tsujimoto, N.; Li, E.; Sasaki, H. Essential role for de novo DNA methyltransferase Dnmt3a in paternal and maternal imprinting. *Nature* **2004**, *429*, 900–903. [CrossRef] [PubMed]
104. Ciccone, D.N.; Su, H.; Hevi, S.; Gay, F.; Lei, H.; Bajko, J.; Xu, G.; Li, E.; Chen, T. KDM1B is a histone H3K4 demethylase required to establish maternal genomic imprints. *Nature* **2009**, *461*, 415–418. [CrossRef] [PubMed]
105. Stewart, K.R.; Veselovska, L.; Kim, J.; Huang, J.; Saadeh, H.; Tomizawa, S.; Smallwood, S.A.; Chen, T.; Kelsey, G. Dynamic changes in histone modifications precede de novo DNA methylation in oocytes. *Genes Dev.* **2015**, *29*, 2449–2462. [CrossRef] [PubMed]
106. Vlachogiannis, G.; Niederhuth, C.E.; Tuna, S.; Stathopoulou, A.; Viiri, K.; de Rooij, D.G.; Jenner, R.G.; Schmitz, R.J.; Ooi, S.K. The Dnmt3L ADD Domain Controls Cytosine Methylation Establishment during Spermatogenesis. *Cell Rep.* **2015**. [CrossRef] [PubMed]
107. Chotalia, M.; Smallwood, S.A.; Ruf, N.; Dawson, C.; Lucifero, D.; Frontera, M.; James, K.; Dean, W.; Kelsey, G. Transcription is required for establishment of germline methylation marks at imprinted genes. *Genes Dev.* **2009**, *23*, 105–117. [CrossRef]
108. Mayer, W.; Niveleau, A.; Walter, J.; Fundele, R.; Haaf, T. Demethylation of the zygotic paternal genome. *Nature* **2000**, *403*, 501–502. [CrossRef] [PubMed]
109. Oswald, J.; Engemann, S.; Lane, N.; Mayer, W.; Olek, A.; Fundele, R.; Dean, W.; Reik, W.; Walter, J. Active demethylation of the paternal genome in the mouse zygote. *Curr. Biol.* **2000**, *10*, 475–478. [CrossRef]
110. Iqbal, K.; Jin, S.G.; Pfeifer, G.P.; Szabo, P.E. Reprogramming of the paternal genome upon fertilization involves genome-wide oxidation of 5-methylcytosine. *Proc. Natl. Acad. Sci. USA* **2011**, *108*, 3642–3647. [CrossRef]
111. Gu, T.P.; Guo, F.; Yang, H.; Wu, H.P.; Xu, G.F.; Liu, W.; Xie, Z.G.; Shi, L.; He, X.; Jin, S.G.; et al. The role of Tet3 DNA dioxygenase in epigenetic reprogramming by oocytes. *Nature* **2011**, *477*, 606–610. [CrossRef] [PubMed]
112. Inoue, A.; Zhang, Y. Replication-dependent loss of 5-hydroxymethylcytosine in mouse preimplantation embryos. *Science* **2011**, *334*, 194. [CrossRef] [PubMed]
113. Inoue, A.; Shen, L.; Dai, Q.; He, C.; Zhang, Y. Generation and replication-dependent dilution of 5fC and 5caC during mouse preimplantation development. *Cell Res.* **2011**, *21*, 1670–1676. [CrossRef]
114. Nakamura, T.; Liu, Y.J.; Nakashima, H.; Umehara, H.; Inoue, K.; Matoba, S.; Tachibana, M.; Ogura, A.; Shinkai, Y.; Nakano, T. PGC7 binds histone H3K9me2 to protect against conversion of 5mC to 5hmC in early embryos. *Nature* **2012**, *486*, 415–419. [CrossRef]

115. Cardoso, M.C.; Leonhardt, H. DNA methyltransferase is actively retained in the cytoplasm during early development. *J. Cell Biol.* **1999**, *147*, 25–32. [CrossRef] [PubMed]
116. Howell, C.Y.; Bestor, T.H.; Ding, F.; Latham, K.E.; Mertineit, C.; Trasler, J.M.; Chaillet, J.R. Genomic imprinting disrupted by a maternal effect mutation in the Dnmt1 gene. *Cell* **2001**, *104*, 829–838. [CrossRef]
117. Hirasawa, R.; Chiba, H.; Kaneda, M.; Tajima, S.; Li, E.; Jaenisch, R.; Sasaki, H. Maternal and zygotic Dnmt1 are necessary and sufficient for the maintenance of DNA methylation imprints during preimplantation development. *Genes Dev.* **2008**, *22*, 1607–1616. [CrossRef]
118. Quenneville, S.; Verde, G.; Corsinotti, A.; Kapopoulou, A.; Jakobsson, J.; Offner, S.; Baglivo, I.; Pedone, P.V.; Grimaldi, G.; Riccio, A.; et al. In embryonic stem cells, ZFP57/KAP1 recognize a methylated hexanucleotide to affect chromatin and DNA methylation of imprinting control regions. *Mol. Cell* **2011**, *44*, 361–372. [CrossRef]
119. Li, X.; Ito, M.; Zhou, F.; Youngson, N.; Zuo, X.; Leder, P.; Ferguson-Smith, A.C. A maternal-zygotic effect gene, *Zfp57*, maintains both maternal and paternal imprints. *Dev. Cell* **2008**, *15*, 547–557. [CrossRef]
120. Mackay, D.J.; Callaway, J.L.; Marks, S.M.; White, H.E.; Acerini, C.L.; Boonen, S.E.; Dayanikli, P.; Firth, H.V.; Goodship, J.A.; Haemers, A.P.; et al. Hypomethylation of multiple imprinted loci in individuals with transient neonatal diabetes is associated with mutations in ZFP57. *Nat. Genet.* **2008**, *40*, 949–951. [CrossRef]
121. Messerschmidt, D.M.; de Vries, W.; Ito, M.; Solter, D.; Ferguson-Smith, A.; Knowles, B.B. Trim28 is required for epigenetic stability during mouse oocyte to embryo transition. *Science* **2012**, *335*, 1499–1502. [CrossRef] [PubMed]
122. Nakamura, T.; Arai, Y.; Umehara, H.; Masuhara, M.; Kimura, T.; Taniguchi, H.; Sekimoto, T.; Ikawa, M.; Yoneda, Y.; Okabe, M.; et al. PGC7/Stella protects against DNA demethylation in early embryogenesis. *Nat. Cell Biol.* **2007**, *9*, 64–71. [CrossRef] [PubMed]
123. Auclair, G.; Guibert, S.; Bender, A.; Weber, M. Ontogeny of CpG island methylation and specificity of DNMT3 methyltransferases during embryonic development in the mouse. *Genome Biol.* **2014**, *15*, 545. [CrossRef] [PubMed]
124. Borgel, J.; Guibert, S.; Li, Y.; Chiba, H.; Schubeler, D.; Sasaki, H.; Forne, T.; Weber, M. Targets and dynamics of promoter DNA methylation during early mouse development. *Nat. Genet.* **2010**, *42*, 1093–1100. [CrossRef] [PubMed]
125. Chen, T. Mechanistic and functional links between histone methylation and DNA methylation. *Prog. Mol. Biol. Transl. Sci.* **2011**, *101*, 335–348. [PubMed]
126. Du, J.; Johnson, L.M.; Jacobsen, S.E.; Patel, D.J. DNA methylation pathways and their crosstalk with histone methylation. *Nat. Rev. Mol. Cell Biol.* **2015**, *16*, 519–532. [CrossRef]
127. Guenatri, M.; Duffie, R.; Iranzo, J.; Fauque, P.; Bourc'his, D. Plasticity in Dnmt3L-dependent and -independent modes of de novo methylation in the developing mouse embryo. *Development* **2013**, *140*, 562–572. [CrossRef] [PubMed]

© 2019 by the authors. Licensee MDPI, Basel, Switzerland. This article is an open access article distributed under the terms and conditions of the Creative Commons Attribution (CC BY) license (http://creativecommons.org/licenses/by/4.0/).

Review

The Roles of Human DNA Methyltransferases and Their Isoforms in Shaping the Epigenome

Hemant Gujar [1], Daniel J. Weisenberger [2] and Gangning Liang [1,*]

1. Department of Urology, Keck School of Medicine, University of Southern California, Los Angeles, CA 90033, USA; gujar@usc.edu
2. Department of Biochemistry & Molecular Medicine, Keck School of Medicine, University of Southern California, Los Angeles, CA 90033, USA; dan.weisenberger@med.usc.edu
* Correspondence: gliang@usc.edu

Received: 7 January 2019; Accepted: 19 February 2019; Published: 23 February 2019

Abstract: A DNA sequence is the hard copy of the human genome and it is a driving force in determining the physiological processes in an organism. Concurrently, the chemical modification of the genome and its related histone proteins is dynamically involved in regulating physiological processes and diseases, which overall constitutes the epigenome network. Among the various forms of epigenetic modifications, DNA methylation at the C-5 position of cytosine in the cytosine–guanine (CpG) dinucleotide is one of the most well studied epigenetic modifications. DNA methyltransferases (DNMTs) are a family of enzymes involved in generating and maintaining CpG methylation across the genome. In mammalian systems, DNA methylation is performed by DNMT1 and DNMT3s (DNMT3A and 3B). DNMT1 is predominantly involved in the maintenance of DNA methylation during cell division, while DNMT3s are involved in establishing *de novo* cytosine methylation and maintenance in both embryonic and somatic cells. In general, all DNMTs require accessory proteins, such as ubiquitin-like containing plant homeodomain (PHD) and really interesting new gene (RING) finger domain 1 (UHRF1) or DNMT3-like (DNMT3L), for their biological function. This review mainly focuses on the role of DNMT3B and its isoforms in *de novo* methylation and maintenance of DNA methylation, especially with respect to their role as an accessory protein.

Keywords: DNA methylation; DNA methyltransferase; DNMT; DNMT3B; epigenetics

---

1. Introduction

With the improvement in DNA-sequencing technologies, genetic alterations were initially recognized as the main contributors of the initiation and progression of diseases including cancer [1–4]. However, epigenetic mechanisms are becoming increasingly recognized as the drivers of tumorigenesis and tumor progression [5–8]. The epigenome of an organism represents the collection of cellular processes that alter gene regulation, which are not attributed to the DNA sequence alone. This information is represented by histone modification, histone variants, nucleosome occupancy, expression of non-coding regulatory RNAs, and DNA methylation. These modifications dynamically regulate gene expression to maintain physiological processes, cellular function, and organismal development [5,7,9–19].

1.1. Nucleosomal Structure and Histone Modifications

Epigenetic and genetic factors interplay to maintain physiological states of cells and are often altered and dysregulated across virtually every form of human cancer [5,6,20,21]. The genome in a cell is tightly packaged around histone proteins to form a basic nucleosome structure [19,22]. Specifically, the nucleosome consists of 145–147 bp of DNA wrapped around a core histone (H) octamer that is composed of two copies each of H2A, H2B, H3, and H4 proteins with a 10–70-bp linker DNA and linker histone H1 attached. As a result, the nucleosomal structure can physically restrict the access of proteins with DNA. Nucleosome occupancy at any given genomic location is defined as the cell population fraction occupied by histones, and may be effected due to various factors such as DNA motifs, DNA methylation, and histone modifications [23]. In general, active functional elements, such as promoters and enhancers, are free of nucleosomes in an open structure often referred to as a nucleosome-depleted region (NDR).

Chemical modifications to chromatin such as histone acetylation, methylation, phosphorylation, ubiquitinylation, sumoylation, adenosine diphosphate (ADP) ribosylation, deamination, propionylation, and butyrylation are known to influence the interaction of histones with non-histone proteins, as well as overall chromatin structure [24] and protein accessibility, which in turn influence chromatin accessibility and regulate gene expression [19,22]. For example, histone H3 lysine 4 trimethylation (H3K4me3) occupancy in gene promoter regions and histone H3 lysine 36 trimethylation (H3K36me3) in gene body regions are both associated with actively transcribed genes, while histone H3 lysine 27 trimethylation (H3K27me3) at gene promoters is associated with suppressed gene expression [22]. Histone reader enzymes scan the nucleosomal topography and structure for recognizing specific histone modifications. Histone tail marks are placed by specific enzymes (writers) and are removed by enzymes known as erasers. A comprehensive list of histone erasers, writers, and readers and the roles of specific histone tail modifications was previously described [6,22,25,26].

The highly conserved core histones are replaced by variant versions as a result of DNA replication and DNA repair mechanisms. Histone variants are expressed throughout the cell cycle with a unique copy number in the genome, and show varying interaction with chromatin modifiers or may simply act as replacements for core histones [27]. The fundamental chromatin structure nucleosome is also regulated by adenosine triphosphate (ATP)-dependent chromatin-remodeling complexes that regulate DNA packaging in the nucleus and, thus, affect the accessibility of various factors to specific DNA regions to regulate gene expression, replication, repair, and recombination [28].

1.2. Non-Coding RNAs

Non-coding RNAs also have important roles in regulating gene expression, signaling networks, and the epigenome [29]. Non-coding RNAs are divided in two groups: small RNAs and long non-coding RNAs (lncRNAs). Small RNAs include microRNAs (miRNAs), small interfering RNAs (siRNAs), and piwi-interacting RNAs (piRNAs), and regulate gene expression through the canonical RNA interference (RNAi) pathway [29,30]. The function of piRNAs is in concert with the piwi protein Aubergine to silence repetitive elements during germ cell development, most notably in *Drosophila* and mouse systems [30–32]. Piwi-interacting RNAs may also act as cancer biomarkers, and P-element-induced wimpy testis (PIWI) proteins are upregulated in several cancer types as compared to normal cells [33]. Long ncRNAs have roles in transcriptional activation and repression, recruiting lineage-specific complexes, serving as scaffolds, regulating RNA splicing, and sequestering miRNAs [34,35]. A well-known lncRNA is *X inactive specific transcript* (*XIST*), RNA that interacts with several proteins to silence the X-chromosome [36]. *XIST* also mediates localization of the H3K27me3 repressive histone mark though the polycomb repressive complex 2 (PRC2) and Jumonji and AT-rich interaction domain containing 2 (JARID2) [37].

1.3. DNA Methylation

Among the various DNA modifications, 5-methylcytosine (5mC) in the 5'-CpG-3' is common and well documented in mammalian genomes [38,39]. DNA methylation influences cellular differentiation, physiological conditions, X-chromosome inactivation [40,41], gene imprinting [42], and repression of retrotransposons [43,44]. DNA methylation profiles are stably inherited in mitotically dividing cells and are conserved with respect to the cell type. During development, DNA demethylation and remethylation occur in the migratory and post-migratory primordial germ cells, as well as in pre-implantation and early post-implantation stages [45].

Cytosine methylation is inherently mutagenic, as DNA methylation drives mutation rates that occur in cancers and other diseases [5,6]. Indeed, 5-methylcytosine is prone to spontaneous deamination to thymine; thus, the CpG sequence context is reduced to approximately 20% of what is predicted in the human genome. Furthermore, 5mC deamination also results in the creation of guanine–thymine mismatches that are repaired in an error-prone fashion, while cytosine-to-uracil deamination is easily flagged for repair, as uracil is not part of the DNA genome code. Methylated cytosines occur predominantly in CpG-poor regions of the genome and at repetitive elements; however, CpG-dense regions, termed CpG islands, have the expected CpG content and are usually devoid of methylation in normal somatic cells [46].

In gene promoter regions, DNA methylation can affect gene expression by regulating the recruitment of methylated DNA binding proteins (MBDs), which influence transcription factor binding and overall chromatin structure [47,48]. Specifically, DNA methylation blocks the interaction of some transcriptional activators with DNA sites or allow binding of repressive factors and insulators containing methyl-CpG-binding proteins to repress transcription [49,50]. In addition, genes marked with H3K27me3 are maintained in the differentiating cells in the presence of EZH2, an H3K27me3 methyltransferase, a component of the PRC2. In cancer cells, this complex recruits DNA methyltransferases (DNMTs) to establish DNA methylation patterns [51].

Concurrently, DNA methylation in gene bodies is associated with actively transcribed genes [48,52,53] and regulation of splicing [54,55]. DNA methylation at transcribed regions stabilizes the nucleosome for efficient RNA polymerase II (Pol II) transcriptional elongation, inhibits spurious transcription, represses retrotransposon activation, and helps facilitate RNA splicing [53,56,57]. In addition, gene body DNA methylation in actively transcribed genes is also associated with the active H3K36me3 histone mark that interacts with elongation factors for high transcription activity [58].

1.4. DNA Methylation Aberrancies in Human Cancers

Genes associated with regulation of the epigenome are often mutated in several tumor types [23]. Epigenetic gene regulation is thought to be a main and early driver in tumorigenesis, especially global DNA hypomethylation and gene promoter DNA hypermethylation [59]. For example, DNA hypermethylation in the promoter regions of the *SFRP* (secreted frizzled-related proteins) tumor suppressor gene family, inhibitors of the WNT signaling pathway, complements the effect of downstream mutations in human colorectal cancer [60]. In addition, *HIC1* (hypermethylated in cancer 1) promoter DNA hypermethylation is an early event in several cancers such as epithelial cancers in males and lymphomas and sarcomas in females [61]. Moreover, promoter DNA hypermethylation of genes involved in DNA repair, such as *MLH1*, *BRCA1*, *BRCA2*, and *XRCC5* correlates with reduced expression and results in enrichment of carcinogenic mutations in several forms of human cancers [62].

1.5. DNA Demethylation

DNA methylation is reversed by the ten-eleven translocase (TET) group of enzymes using vitamin C as a co-factor. TET enzymes oxidize 5-methyl-cytosine to cytosine through a 5-hydroxymethyl-cytosine (5-hmC) intermediate. The oxidation of 5-hmC generates 5-formylcytosine and 5-carboxycytosine, which are then replaced with unmethylated cytosine through the base excision repair pathway [63]. TET proteins have a common C-terminal catalytic domain. TET1 and TET3 have an additional CXXC domain which may facilitate binding to DNA [63]. TET enzymes are found at a higher concentration at transcription start sites of CpG island promoter regions [64]. TET mutations were reported in several solid tumors and myeloid tumors where their loss of function might promote aberrant DNA methylation [65]. TET2 mutations affecting enzyme function were reported in myeloid disorders, whereas TET1 and TET3 mutations were also in hematopoietic malignancies [63]. In addition, low levels of 5-hmC were observed in various cancer types, suggesting that low TET activity contributes to tumorigenesis and cancer progression [66].

2. DNA Methyltransferases (DNMTs)

DNA methylation is catalyzed by a conserved set of enzymes, termed DNMTs. DNMTs contain an N-terminal regulatory domain and a C-terminal catalytic domain, and include DNMT1, DNMT3A, DNMT3B, and Dnmt3c. The structure, mechanism, and function of DNMTs were previously described in detail [67–71]. An unusual DNMT reported is DNMT2, which consists of only a catalytic domain. Subsequently, DNMT2 was shown to lack DNA methyltransferase activity and functions instead as a transfer RNA (tRNA) methyltransferase [72]. DNMTs are highly expressed during DNA replication in the synthesis S-phase of the cell cycle, as well as in embryonic, somatic, and cancer cell types [73].

DNMTs function in maintenance and *de novo* functions, with the DNA methylation profiles generally maintained after DNA replication and cell division by DNMT1, while DNMT3A and DNMT3B are largely involved in facilitating *de novo* DNA methylation profiles at loci that were previously unmethylated [74]. However, it should be noted that DNMT1, DNMT3A, and DNMT3B function together in maintaining DNA methylation during DNA replication. In vitro studies also indicated a role of Dnmt1 in *de novo* methylation maintenance succeeding the Dnmt3 activity in murine systems [75]. DNMT1 is the primary methyltransferase its activity is sufficient to methylate DNA in CpG-poor regions [76,77]. However, CpG-dense regions require cooperativity with DNMT3A/3B to maintain DNA methylation. In addition, repetitive elements require both Dnmt1 and Dnmt3a and/or Dnmt3b for maintenance of DNA methylation in mouse embryonic stem cells [76].

DNMT3A and DNMT3B are mainly expressed in undifferentiated cells, where they are essential for the formation and subsequent maintenance of DNA methylation marks [15,76,78,79]. Both DNMTs are expressed in several isoforms that are essential during embryonic development and are dysregulated in human cancers [80–84]. Although DNMT1 is a key enzyme for maintenance of DNA methylation, the double knockout of *Dnmt3a* and *Dnmt3b* in mouse embryonic cells resulted in a gradual loss of DNA methylation over time, indicating the involvement of Dnmt3a/Dnmt3b in maintaining DNA methylation profiles during embryonic development (Figure 1) [76,77].

In general, DNMT3A or DNMT3B have roles in hematopoietic stem cell (HSC) self-renewal, maturation, and differentiation [85]. Knocking out *Dnmt3b* in mouse embryonic cells revealed its roles in bone mineralization, normal limb development, and bone morphogenetic protein signaling [86], while Dnmt3a affects bone density by effecting osteoclast differentiation [87]. Dnmt3b is also required postnatally in mice for fracture repair and angiogenesis [88], and human DNMT3B mutations are known to cause immunodeficiency, centromeric instability, and facial anomalies (ICF) syndrome [83], in which loss of DNMT3B activity results in DNA hypomethylation of the centromeric regions [89], as well as reduced interaction with the DNMT3L accessory protein [81].

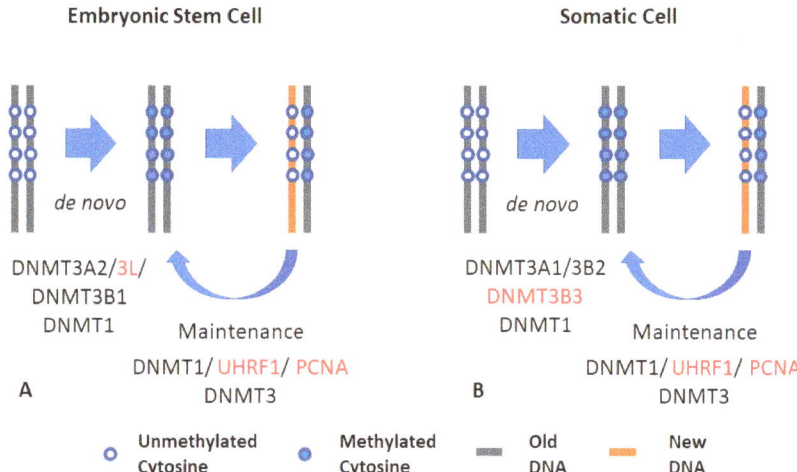

Figure 1. General concept of DNA methylation: (**A**) DNA methylation in embryonic stem cells (ESCs): ESCs express DNA methyltransferases 3A2 and 3L (DNMT3A2 and DNMT3L), which interact with each other through the carboxy domain. DNMT3L interacts with the unmethylated form of histone H3K4. The catalytically active isoform DNMT3B1 is also highly expressed in embryonic cells. UHRF1 and DNMT1 also maintain DNA methylation in ESCs. (**B**) DNA methylation in somatic cells: somatic cells mainly express and maintain methylation using DNMT1 and its accessory protein UHRF1. Somatic cells express the catalytically active DNMT3A1 and DNMT3B2 (at low expression levels) that interact with DNMT3B isoforms to catalyze methyl transfer. This interaction is required in the maintenance process. Somatic cells also express the catalytically inactive isoform, DNMT3B3. The black font indicates active DNMT, while the red font indicates an accessary protein.

2.1. DNMT1

DNMT1 is mainly responsible for the maintenance of DNA methylation and performs this task by copying the DNA methylation patterns from the parental DNA strand to the daughter DNA strand in coordination with DNA replication and cell division. DNMT1 consists of a C-terminal catalytic domain and an N-terminal regulatory domain. Interestingly, the DNMT1 C-terminal domain consists of two sub-domains: the methyltransferase domain and the target-recognition domain (TRD) responsible for recognizing hemimethylated cytosines [90]. The regulatory domain contains a DNA methyltransferase 1-associated protein 1 (DMAP1) binding domain, a replication foci targeting sequence (RFTS) domain, a CXXC domain, and a tandem bromo-adjacent homology (BAH1/2) domain [91]. *De novo* activity of DNMT1 is inhibited by the BAH domain, which positions itself between the unmethylated CpG-binding CXXC domain and the catalytic site [70]. Alternatively, the RFTS and CXXC domains may also interact with each other to inhibit DNMT1 access to DNA [92].

The high fidelity of DNMT1 toward hemimethylated DNA is maintained by both self-regulation of DNMT1 activity and by the accessory protein UHRF1 [93]. UHRF1 binds to DNA using its methyl DNA binding domain SRA (SET and RING associated), which preferentially binds to hemimethylated DNA [93]. UHRF1 directly recruits DNMT1 to hemimethylated DNA by removing the auto-inhibitory mechanism of the RFTS domain [71,92]. Interaction of proliferating cell nuclear antigen (PCNA), an accessory protein in DNA replication and repair, also reportedly interacts with DNMT1 [94]. In mouse embryonic stem cells (ESCs), the interaction of PCNA with DNMT1 enhances the efficiency and fidelity of the maintenance of DNA methylation [95].

Loss of Dnmt1 activity was shown to cause neurological abnormalities [96]. Loss of Dnmt1 in mouse embryonic cells leads to genome-wide demethylation, confirming the importance of DNMT1 in maintenance methylation [97]. Similarly, in mammalian systems, *DNMT1* or *UHRF1* knockout leads to dramatic global DNA hypomethylation [93,96,98]. In HCT116 colon cancer cells, loss of DNMT1 results in cell-cycle arrest and mitotic defects, leading to cell death [99]. Interestingly, DNMTs are often aberrantly expressed in several cancer types, and normal DNMT expression is essential for development in mammals [100]. Similarly, UHRF1 expression is also upregulated in cancer, resulting in global DNA hypomethylation [101,102]. Recent studies found the expression of two variant Dnmt1 isoforms: (1) Dnmt1o, which is expressed in the oocyte and pre-implantation embryo, and (2) Dnmt1p, which is expressed in pachytene spermatocytes [103,104]. Both Dnmt1o and Dnmt1p are expressed as a result of alternate usage of the first exon, and their loss results in abnormal DNA methylation of imprinted genes and lethality in the developing fetus [104].

2.2. DNMT3A

The *DNMT3A* locus expresses two isoforms, DNMT3A1 and DNMT3A2. *DNMT3A2* is expressed from an intronic promoter downstream of the *DNMT3A1* promoter, and lacks 223 amino acids in the N-terminal region. Both isoforms contain the PWWP domain for H3K36me3 interaction and the ADD domain for histone binding and transcriptional and epigenetic regulation. DNMT3A1 expression is maintained in differentiated cells and is localized to heterochromatic regions, where it interacts with DNA using the N-terminal domain [84,105]. In mice, undifferentiated cells, embryonic stem cells, and embryonal carcinoma cells, as well as the testis, ovary, thymus, and spleen, predominantly express Dnmt3a2 [106]. DNMT3A2 is the predominant DNMT3A isoform responsible for *de novo* DNA methylation in embryonic stem cells (Figure 1). DNMT3A2 expression is localized at euchromatic regions of the genome [106] where it is recruited by DNMT3L through its C-terminal catalytic domain.

DNMT3A mutations result in dysregulation of its activity and aberrant DNA methylation patterns commonly reported during various developmental defects and hematological malignancies among others. Mouse *Dnmt3a* knockout (*Dnmt3a* −/−) nervous tissues resulted in retarded neuromuscular and motor activity, and subsequent premature death in adults; however, the embryos and birth rate were seemingly not affected [107]. DNMT3A mutations were also reported in overgrowth disorders in humans [108]. The symptoms of this syndrome include facial abnormality, pre- and postnatal overgrowth, and intellectual disability. DNMT3A mutations are also frequent in acute myeloid leukemia (AML) patients. The R882 (arginine) residue is the most frequent frame-shift mutation in AML patients; however, nonsense and splice site mutations were also reported. The R882H is a dominant negative mutation in the catalytic domain of DNMT3A, which affects the activity of the wild-type enzyme by inhibiting its homotetramerization [109].

2.3. DNMT3B

DNMT3B is expressed in more than 30 isoforms [110–112] with alternative splicing occurring in both the catalytic and regulatory domains. Full-length DNMT3B, DNMT3B1, is highly expressed in embryonic stem cells, but its expression decreases in somatic cells. DNMT3B3 is noted by alternative splicing of the catalytic domain, which was shown to affect its methyl-transfer function. DNMT3B3 is highly expressed in somatic cells (Figure 1) [113]. Interestingly, several catalytically inactive and active isoforms of DNMT3B were also discovered to act as accessory proteins in coordinating DNA methylation [113].

2.4. Dnmt3c

Recently, a duplicated copy of Dnmt3b, specific to mouse, termed Dnmt3c, was documented. Dnmt3c is truncated at the N-terminal domain and lacks the PWWP domain. Dnmt3c is exclusively expressed in male gonads and is involved in promoter DNA methylation of evolutionarily young retrotransposons. Dnmt3c resulted in male sterility, while females were unaffected [114].

2.5. DNMT3L

Dnmt3L acts as an accessory protein for Dnmt3a- and Dnmt3b-mediated *de novo* DNA methylation during gametogenesis and in embryos soon after fertilization [115]. Dnmt3L is expressed only in germ cells and embryonic stem cells, but not in somatic cells [116], and is involved in the regulation of repetitive elements and imprinting in germ cells [115,117]. Dnmt3L is predominantly expressed in the postnatal female germline along with Dnmt3a and Dnmt3b for the establishment of DNA methylation patterns. In contrast, Dnmt3L and Dnmt3a are expressed in male prenatal tissues [118–120].

The PHD domain of Dnmt3L interacts with genomic regions marked by unmethylated H3K4 [121]. This interaction also stabilizes and regulates the activity of Dnmt3a2 in mouse embryonic cells [122]. Mass spectroscopy of epitope-tagged Dnmt3L in mice revealed its interaction with Dnmt3a, Dnmt3b, and core histones, in that the *N*-terminal cysteine rich domain of Dnmt3L interacts with histone H3, and this interaction is inhibited in the presence of H3K4me3 [121]. In mouse embryonic stem cells, Dnmt3L regulates gene expression by promoting gene body DNA methylation of housekeeping genes. Contrary to popular belief, this group also showed that Dnmt3L negatively regulates DNA methylation at bivalent promoters (carrying both the active H3K4me3 histone mark and the repressive H3K27me3 histone mark) [123]. This is achieved through a competition between PRC2 and Dnmt3 to bind Dnmt3L at H3K27me3-occupied loci. [123]. More recently, these results were challenged by similar studies using *Dnmt3l* knockout cells from blastocyst-stage mouse embryos, where Dnmt3L was shown to contribute to Dnmt3a-dependent DNA methylation [122].

DNMT3L knockout human embryonal carcinoma cells resulted in apoptosis and suppression of growth [124]. In human fetal-derived kidney cells, co-expression of both DNMT3L and DNMT3A, but not DNMT3B, is required for DNA methylation of maternally imprinted genes [125]. The lack of DNMT3L expression in somatic cells raises an interesting point as to the unidentified accessory protein(s) during DNMT3A/DNMT3B-mediated DNA methylation in somatic cells. Recently, the catalytically inactive DNMT3B3 and DNMT3B4 isoforms and the catalytically active DNMT3B1 isoform were shown to acts as accessory proteins to DNMT3A/3B function [113,126].

3. Specific Roles of DNMT3B Isoforms

In addition to their *de novo* methyltransferase function, DNMT3A and 3B have affinity toward methylated CpG sites in the genome. A recent study from our group demonstrated that DNA hypermethylation at transcribed gene regions is correlated with upregulated gene expression, and demethylating these hypermethylated regions via the DNA methylation inhibitor 5-aza-2′-deoxycytidine (5-aza-CdR) also downregulated the expression of these genes [48]. *DNMT3B* knockout cells resulted in reduced DNA remethylation after 5-aza-CdR treatment, highlighting its function in gene body DNA methylation, whereas *DNMT1* knockout had no effect on gene body DNA methylation. In addition, gene body DNA methylation was also correlated with the presence of H3K36me3 occupancy (Figure 2) [48,127], suggesting that DNMT3B may interact with H3K36me3 marks or histone-lysine N-methyltransferase SET domain containing 2 (SETD2), the enzyme responsible for placing these marks. To prove the role of H3K36me3 in recruiting DNMT3B, H3K36me3 levels were reduced by *SETD2* knockout. Depleted H3K36me3 was correlated with reduced DNMT3 co-localization. The unbound DNMT3A and DNMT3B are rapidly degraded through proteasomal and other unknown pathways [128]. In addition, the interaction between H3K36me3 and the PWWP domain of DNMT3B1 was also confirmed through in vitro experiments [78]. The chemical inhibition of RNA polymerase II did not affect DNMT3B binding to H3K36me3, showing its independence to transcription activity [78].

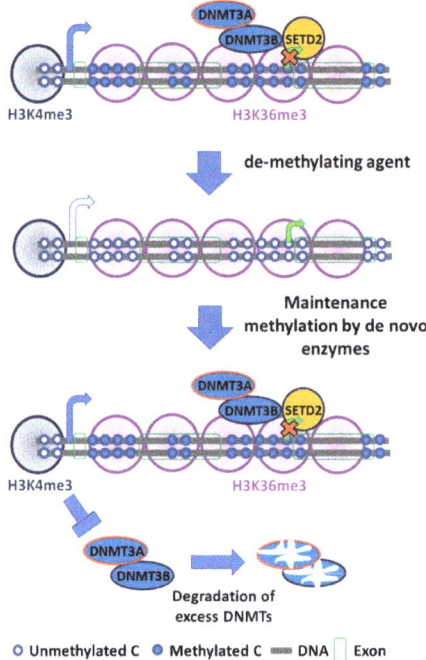

Figure 2. Mechanism of *de novo* DNA methyltransferases (DNMTs): DNMT3s localizes on methylated cytosine–guanine (CpG)-rich locus to maintain a uniform methylation pattern. DNMT3B targets gene body to methylate cytosine locus. DNMT3B acts as maintenance enzyme, complimenting the low fidelity of DNMT1 and as a *de novo* enzyme to mark new methylation patterns during differentiation and remethylation after treatment with demethylating agents. DNMT3B also interact with H3K36me3 to localize at these active expression marks. DNA methylation in the genic region is essential for efficient transcription, stability of splicing factors, stability of elongation factor, and to inhibit generation of spurious transcripts. DNMT3s localization at the promoter region is prevented by unmethylated cytosines and by H3K4me3. The free DNMT3s are unstable and are degraded.

As the embryonic stem cells differentiate, DNMT3B1 expression (catalytically active) decreases and is replaced by increased expression of the catalytically inactive DNMT3B3 isoform. Since DNMT3B is responsible for transcribed gene body DNA methylation and DNA remethylation after 5-aza-CdR treatment, *DNMT3B* knockout cells fail to remethylate gene bodies after 5-aza-CdR treatment [113]. However, DNA remethylation can be restored in *DNMT3B* knockout cells if expression of DNMT3B1, DNMT3B3 (catalytically inactive), DNMT3B1M (mutated catalytic domain), or DNMT3L is restored. Furthermore, *DNMT3A* and *DNMT3B* double knockout cells including the inactive DNMT isoforms (DNMT3B1-M, DNMT3B3, and DNMT3L) fail to restore DNA methylation after 5-aza-CdR treatment. Thus, DNMT3B3 (without catalytic domain), DNMT3B1M (mutated catalytic domain), and DNMT 3L can remethylate CpG loci only in the presence of DNMT3A. Interestingly, gene body DNA remethylation is rescued after 5-aza-CdR treatment when the catalytically active DNMT3B1 isoform is exogenously expressed in *DNMT3A/DNMT3B* double knockout cells, demonstrating its involvement in gene body DNA methylation without DNMT3A involvement [113]. Thus, DNMT3B isoforms with or without its catalytic activity can act as accessory proteins for DNA methylation by recruiting DNMT3A to target regions [113,126]. This experiment also showed that DNMT3B1 may act as the main catalytic and accessory enzyme in gene body DNA methylation in somatic cells. It also showed that some catalytically inactive isoforms of DNMT3B may act as accessory proteins in *de novo* DNA methylation mechanisms. Somatic cells also express the catalytically active DNMT3B2 isoform that contains an intact catalytic domain but lacks exon 10 [74,129]. DNMT3B2 is expressed at low levels in somatic cells [130]. Thus, the characterization of all DNMT3B isoforms in terms of their unique and overlapping accessory roles may also explain the presence of aberrant DNA methylation patterns during tumorigenesis, and may highlight their role as a driving force for the *de novo* DNA methylation profiles in human cancers [113].

In accordance to the aforementioned absence of DNMT3B1 expression in somatic tissues, DNMT3B1 was found to be undetectable in normal lung tissue, while it was highly expressed in non-small-cell lung cancer tissues [111]. Recently, the catalytically inactive DNMT3B3 isoform was demonstrated to act as an accessory protein, much like DNMT3L, in cancer cells [113]. In addition, in vitro DNMT3B3 overexpression counteracted the stimulatory effect of DNMT3L by complexing with DNMT3L, resulting in reduced DNA methylation [126]. These results indicate that the aberrant expression of DNMTs in various forms of human cancer may disrupt DNA methylation profiles.

There are more than 30 DNMT3B isoforms [113,131,132] in human and mouse cells. The domain structures of the most common isoforms are presented in Figure 3. Most of these isoforms are aberrantly expressed in cancer cells but not in normal cells [110]. These isoforms are characterized by exon deletion, premature 3′ termination, alternatively spliced exons, fusion of intronic sequences, and missing 5′ exons. These modifications affect their catalytic activity, as well as their cellular localization and interactions with DNA and the nucleosome, [133]. The specific functional roles of most isoforms are not known. However, the roles of some isoforms were reported and are discussed here. Deletion of part of or the entire catalytic domain, located in the C-terminal region of DNMT3B isoforms, is characteristic of the DNMT3B3, 3B4, 3B5, 3B6, and 3B7 isoforms, all of which are catalytically inactive. DNMT3B4, B5, and B7 have an additional frame-shift mutation that introduces an early stop codon, thereby causing inactivity of the enzymes. Interestingly DNMT3B4 failed to remethylate DNA in the presence of DNMT3A which suggested that the loss of catalytic domain of DNMT3B4 also affected its ability to act as an accessory protein [113], this may suggest a failure of DNMT3B4 to bind to nucleosome. The specific functions and mechanisms of action of these novel exons coded by these frame shifts remain unknown. Deletions in the DNMT3B C-terminal region may affect its interaction with DNMT3L, as in vitro studies showed that the DNMT3B1 C-terminal catalytic domain interacts with DNMT3L [133].

Overexpression of the catalytically inactive DNMT3B isoforms is associated with aberrant DNA methylation patterns. In hepatocellular carcinoma, DNMT3B4 overexpression is associated with hypomethylation of pericentromeric satellite regions [134]. This might be due to the competition of DNMT3B4 with the catalytically active DNMT3B3 isoform that is expressed in normal liver cells [134]. More recently, the sequestering nature of DNMT3B4 toward DNMT3A/3B was also reported, and may be a contributing factor of pericentromeric satellite DNA hypomethylation and genomic instability [126]. The DNMT3B7 isoform may also contribute aberrant alteration in DNA methylation and gene expression in cancer [110], as it is widely expressed in various hematopoietic and solid tumor cell lines. In addition, DNMT3B7 was also found to be highly expressed in differentiated ganglioneuroblastomas as compared to the undifferentiated neuroblastomas. DNMT3B7 expression in neuroblastoma cells correlates with altered gene expression and tumor growth inhibition [135].

DNMT3B isoforms lacking the N-terminal domain, termed ΔDNMT3B isoforms, first reported by Wang et al., are predominantly expressed in non-small-cell lung cancer (NSCLC) [111]. Seven variants of those isoforms were reported and, interestingly, ΔDNMT3B1, ΔDNMT3B2, and ΔDNMT3B4 expression was correlated with *CDKN2A (p16)* and *RASSF1A* promoter DNA methylation in tumor cells [136,137]. The N-terminal region of DNMT3B is essential for strong nucleosome binding, as shown by experiments on Δisoforms of DNMT3B2 (ΔDNMT3B2) that lack the N-terminal domain, as well as ΔDNMT3B4 that lacks both the N-terminal and the PWWP domains [79,128]. The weak association of ΔDNMT3B with nucleosomes eventually results in decreased DNA methylation [113,133]. In vitro studies in which the PHD-like or PWWP domains and the region joining the PWWP and PHD domains were deleted did not have any effect on the interaction between DNMT3B and DNMT3L, whereas deletion of the N-terminal region and a portion of the PWWP domain inhibited interaction with DNMT3B1 [133]. The PWWP domains of DNMT3s are involved in interactions with DNA, and their absence reduces DNMT3 binding affinity to DNA [138]. Furthermore, the PWWP domain modulates interaction between epigenetic marks, together with the histones and DNA modifiers or readers [139]. These findings indicated that aberrantly expressed DNMT3B isoforms in cancer may have specific DNA target sequences, which was later confirmed [131].

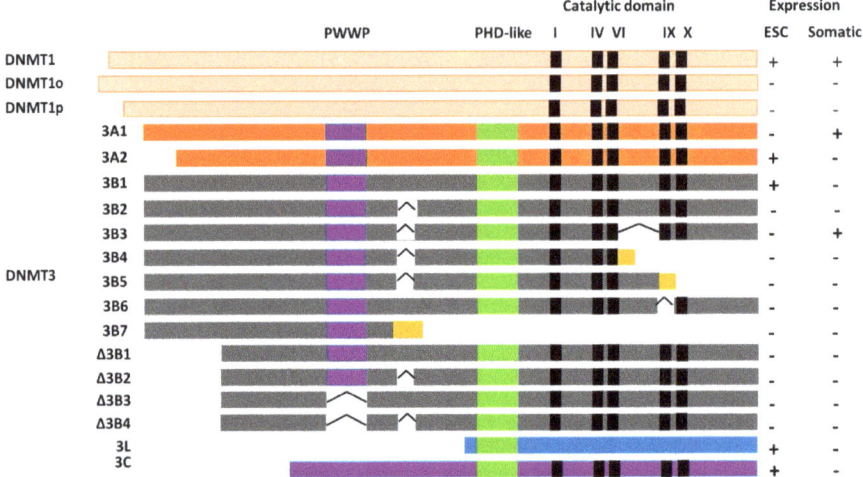

Figure 3. Schematic diagram showing DNA methyltransferase (DNMT) isoforms: DNMT3A (A1 and A2), DNMT3B (3B1, 3B2, 3B3, 3B4, 3B5, 3B6, 3B7, Δ3B1, Δ3B2, Δ3B3, and Δ3B4), and DNMT3L. DNMT3 consist of a PWWP domain (purple), a PHD-like domain (green), and a catalytic domain (black). The deletions are shown as black lines; frame-shift mutations are in yellow. The figure was modified and adapted from Duymich et al., 2016 [113]; Ostler et al., 2007 [110]; Gopalakrishnan et al., 2009 [112]; and Choi et al., 2011 [131]. ESC: Embryonic stem cells.

4. Cooperativity Between DNA Methyltransferases

Established DNA methylation profiles are maintained during each cell division; however, the presence of hemimethylated sequences was reported in a very small proportion of single-copy genes and repetitive elements [76]. This proportion increased in *Dnmt3a/Dnmt3b* double knockout cells, indicating cooperativity between DNMTs [77]. However, Dnmt1 does not have an absolute fidelity, and proofreads to methylate unmethylated DNA quickly after cellular replication and division [76,95]. In vitro experiments implicated the proof-reading activity of DNMT1 in *de novo* DNA methylation mediated by DNMT3A. DNA methylation was observed to be five times higher in the presence of DNMT3A and DNMT1 [75]. The cooperativity between Dnmt3s and Dnmt1 was later also confirmed by experiments in cultured mouse embryonic stem cells [140]. An exception exists in the adult intestinal cells, where Dnmt1 or Dnmt3B alone is sufficient to maintain DNA methylation in dividing cells. *Dnmt1* knockout correlates with induced Dnmt3B expression, and the ablation of both Dnmts resulted in a genome-wide DNA hypomethylation, genomic instability, and increased apoptosis and lethality [141].

5. Future Perspectives

Since the initial discovery of epigenetic factors as drivers for diseases including cancer, many tools and strategies were developed for studying the mechanisms contributing to diseases. This led to a seemingly large number of tools useful for the treatment of epigenetic diseases. The discovery of 5-azacytidine (5-aza-CR) and 5-aza-CdR as DNA demethylation agents [142] shows promise in the treatment of several forms of human cancer. Both drugs were approved by the Food and Drug Administration for the treatment of myelodysplastic syndrome (MDS) and acute myeloid leukemia (AML) [143–145]. A more comprehensive list of drugs for targeting the epigenome is provided in Reference [146]. DNMT accessory proteins may also be efficacious targets for epigenetic therapies, as are histone-modifying enzymes that regulate chromatin structure [147]. The development of plasmid constructs carrying a fusion of an altered clustered regularly interspaced short palindromic repeats (CRISPR)–Cas9 system with a DNMT/TET construct may one day be used in curing diseases related to gene imprinting or cancer [148–150]. Nonetheless, unveiling epigenetic enzymes and their mechanisms of action holds tremendous promise for therapeutics and our understanding of cellular processes.

Author Contributions: Conceptualization, G.L.; validation, G.L.; writing—original draft preparation, H.G.; writing—review and editing, H.G., D.J.W., and G.L.; visualization, G.L.; supervision, G.L.; project administration, G.L.; funding acquisition, G.L.

Funding: This research was funded by the Vicky Joseph Cancer Research Foundation, and the National Cancer Institute, Grant Number 5R21CA201865-02 (G.L.), R35CA209859 (G.L), P30 CA014089 (D.J.W.).

Conflicts of Interest: Daniel J. Weisenberger is a consultant for the Zymo Research Corporation.

References

1. Vogelstein, B.; Papadopoulos, N.; Velculescu, V.E.; Zhou, S.; Diaz, L.A., Jr.; Kinzler, K.W. Cancer genome landscapes. *Science* **2013**, *339*, 1546–1558. [CrossRef] [PubMed]
2. Veltman, J.A.; Brunner, H.G. De novo mutations in human genetic disease. *Nat. Rev. Genet.* **2012**, *13*, 565–575. [CrossRef] [PubMed]
3. Bailey, M.H.; Tokheim, C.; Porta-Pardo, E.; Sengupta, S.; Bertrand, D.; Weerasinghe, A.; Colaprico, A.; Wendl, M.C.; Kim, J.; Reardon, B.; et al. Comprehensive characterization of cancer driver genes and mutations. *Cell* **2018**, *174*, 1034–1035. [CrossRef] [PubMed]
4. Watson, I.R.; Takahashi, K.; Futreal, P.A.; Chin, L. Emerging patterns of somatic mutations in cancer. *Nat. Rev. Genet.* **2013**, *14*, 703–718. [CrossRef] [PubMed]
5. You, J.S.; Jones, P.A. Cancer genetics and epigenetics: Two sides of the same coin? *Cancer Cell* **2012**, *22*, 9–20. [CrossRef] [PubMed]
6. Shen, H.; Laird, P.W. Interplay between the cancer genome and epigenome. *Cell* **2013**, *153*, 38–55. [CrossRef] [PubMed]
7. Baylin, S.B.; Jones, P.A. A decade of exploring the cancer epigenome—Biological and translational implications. *Nat. Rev. Cancer* **2011**, *11*, 726–734. [CrossRef] [PubMed]
8. Wainwright, E.N.; Scaffidi, P. Epigenetics and cancer stem cells: Unleashing, hijacking, and restricting cellular plasticity. *Trends Cancer* **2017**, *3*, 372–386. [CrossRef] [PubMed]
9. Allis, C.D.; Jenuwein, T. The molecular hallmarks of epigenetic control. *Nat. Rev. Genet.* **2016**, *17*, 487–500. [CrossRef] [PubMed]
10. Chen, T.; Dent, S.Y. Chromatin modifiers and remodellers: Regulators of cellular differentiation. *Nat. Rev. Genet.* **2014**, *15*, 93–106. [CrossRef] [PubMed]
11. Fyodorov, D.V.; Zhou, B.R.; Skoultchi, A.I.; Bai, Y. Emerging roles of linker histones in regulating chromatin structure and function. *Nat. Rev. Mol. Cell. Biol.* **2018**, *19*, 192–206. [CrossRef] [PubMed]
12. Jones, P.A. Functions of DNA methylation: Islands, start sites, gene bodies and beyond. *Nat. Rev. Genet.* **2012**, *13*, 484–492. [CrossRef] [PubMed]
13. Jones, P.A.; Baylin, S.B. The fundamental role of epigenetic events in cancer. *Nat. Rev. Genet.* **2002**, *3*, 415–428. [CrossRef] [PubMed]
14. Jones, P.A.; Baylin, S.B. The epigenomics of cancer. *Cell* **2007**, *128*, 683–692. [CrossRef] [PubMed]
15. Jones, P.A.; Liang, G. Rethinking how DNA methylation patterns are maintained. *Nat. Rev. Genet.* **2009**, *10*, 805–811. [CrossRef] [PubMed]
16. Koch, A.; Joosten, S.C.; Feng, Z.; de Ruijter, T.C.; Draht, M.X.; Melotte, V.; Smits, K.M.; Veeck, J.; Herman, J.G.; Van Neste, L.; et al. Analysis of DNA methylation in cancer: Location revisited. *Nat. Rev. Clin. Oncol.* **2018**, *15*, 459–466. [CrossRef] [PubMed]

17. Law, J.A.; Jacobsen, S.E. Establishing, maintaining and modifying DNA methylation patterns in plants and animals. *Nat. Rev. Genet.* **2010**, *11*, 204–220. [CrossRef] [PubMed]
18. Smith, Z.D.; Meissner, A. DNA methylation: Roles in mammalian development. *Nat. Rev. Genet.* **2013**, *14*, 204–220. [CrossRef] [PubMed]
19. Venkatesh, S.; Workman, J.L. Histone exchange, chromatin structure and the regulation of transcription. *Nat. Rev. Mol. Cell. Biol.* **2015**, *16*, 178–189. [CrossRef] [PubMed]
20. Flavahan, W.A.; Gaskell, E.; Bernstein, B.E. Epigenetic plasticity and the hallmarks of cancer. *Science* **2017**, *357*. [CrossRef] [PubMed]
21. Chatterjee, A.; Rodger, E.J.; Eccles, M.R. Epigenetic drivers of tumourigenesis and cancer metastasis. *Semin. Cancer Biol.* **2018**, *51*, 149–159. [CrossRef] [PubMed]
22. Lawrence, M.; Daujat, S.; Schneider, R. Lateral Thinking: How histone modifications regulate gene expression. *Trends Genet.* **2016**, *32*, 42–56. [CrossRef] [PubMed]
23. Andreu-Vieyra, C.V.; Liang, G. Nucleosome occupancy and gene regulation during tumorigenesis. *Adv. Exp. Med. Biol.* **2013**, *754*, 109–134. [CrossRef] [PubMed]
24. Bannister, A.J.; Kouzarides, T. Regulation of chromatin by histone modifications. *Cell. Res.* **2011**, *21*, 381–395. [CrossRef] [PubMed]
25. Falkenberg, K.J.; Johnstone, R.W. Histone deacetylases and their inhibitors in cancer, neurological diseases and immune disorders. *Nat. Rev. Drug Discov.* **2014**, *13*, 673–691. [CrossRef] [PubMed]
26. Feinberg, A.P.; Koldobskiy, M.A.; Gondor, A. Epigenetic modulators, modifiers and mediators in cancer aetiology and progression. *Nat. Rev. Genet.* **2016**, *17*, 284–299. [CrossRef] [PubMed]
27. Talbert, P.B.; Henikoff, S. Histone variants on the move: substrates for chromatin dynamics. *Nat. Rev. Mol. Cell. Bio.* **2017**, *18*, 115–126. [CrossRef] [PubMed]
28. Clapier, C.R.; Iwasa, J.; Cairns, B.R.; Peterson, C.L. Mechanisms of action and regulation of ATP-dependent chromatin-remodelling complexes. *Nat. Rev. Mol. Cell. Biol.* **2017**, *18*, 407–422. [CrossRef] [PubMed]
29. Esteller, M. Non-coding RNAs in human disease. *Nat. Rev. Genet.* **2011**, *12*, 861–874. [CrossRef] [PubMed]
30. Holoch, D.; Moazed, D. RNA-mediated epigenetic regulation of gene expression. *Nat. Rev. Genet.* **2015**, *16*, 71–84. [CrossRef] [PubMed]
31. Carmell, M.A.; Girard, A.; van de Kant, H.J.; Bourc'his, D.; Bestor, T.H.; de Rooij, D.G.; Hannon, G.J. MIWI2 is essential for spermatogenesis and repression of transposons in the mouse male germline. *Dev. Cell.* **2007**, *12*, 503–514. [CrossRef] [PubMed]
32. Kuramochi-Miyagawa, S.; Watanabe, T.; Gotoh, K.; Totoki, Y.; Toyoda, A.; Ikawa, M.; Asada, N.; Kojima, K.; Yamaguchi, Y.; Ijiri, T.W.; et al. DNA methylation of retrotransposon genes is regulated by Piwi family members MILI and MIWI2 in murine fetal testes. *Genes Dev.* **2008**, *22*, 908–917. [CrossRef] [PubMed]
33. Assumpcao, C.B.; Calcagno, D.Q.; Araujo, T.M.; Santos, S.E.; Santos, A.K.; Riggins, G.J.; Burbano, R.R.; Assumpcao, P.P. The role of piRNA and its potential clinical implications in cancer. *Epigenomics* **2015**, *7*, 975–984. [CrossRef] [PubMed]
34. Malik, B.; Feng, F.Y. Long noncoding RNAs in prostate cancer: Overview and clinical implications. *Asian J. Androl.* **2016**, *18*, 568–574. [CrossRef] [PubMed]
35. Lee, J.T. Epigenetic regulation by long noncoding RNAs. *Science* **2012**, *338*, 1435–1439. [CrossRef] [PubMed]
36. McHugh, C.A.; Chen, C.K.; Chow, A.; Surka, C.F.; Tran, C.; McDonel, P.; Pandya-Jones, A.; Blanco, M.; Burghard, C.; Moradian, A.; et al. The *Xist* lncRNA interacts directly with SHARP to silence transcription through HDAC3. *Nature* **2015**, *521*, 232–236. [CrossRef] [PubMed]
37. da Rocha, S.T.; Boeva, V.; Escamilla-Del-Arenal, M.; Ancelin, K.; Granier, C.; Matias, N.R.; Sanulli, S.; Chow, J.; Schulz, E.; Picard, C.; et al. Jarid2 is implicated in the initial *Xist*-induced targeting of PRC2 to the inactive X chromosome. *Mol. Cell* **2014**, *53*, 301–316. [CrossRef] [PubMed]
38. Raiber, E.A.; Hardisty, R.; van Delft, P.; Balasubramanian, S. Mapping and elucidating the function of modified bases in DNA. *Nat. Rev. Chem.* **2017**, *1*. [CrossRef]
39. Breiling, A.; Lyko, F. Epigenetic regulatory functions of DNA modifications: 5-methylcytosine and beyond. *Epigenetics Chromatin.* **2015**, *8*, 24. [CrossRef] [PubMed]
40. Riggs, A.D. X inactivation, differentiation, and DNA methylation. *Cytogenet. Cell. Genet.* **1975**, *14*, 9–25. [CrossRef] [PubMed]

41. Gendrel, A.V.; Apedaile, A.; Coker, H.; Termanis, A.; Zvetkova, I.; Godwin, J.; Tang, Y.A.; Huntley, D.; Montana, G.; Taylor, S.; et al. Smchd1-dependent and -independent pathways determine developmental dynamics of CpG island methylation on the inactive X chromosome. *Dev. Cell.* **2012**, *23*, 265–279. [CrossRef] [PubMed]
42. Henckel, A.; Arnaud, P. Genome-wide identification of new imprinted genes. *Brief. Funct. Genomics* **2010**, *9*, 304–314. [CrossRef] [PubMed]
43. Slotkin, R.K.; Martienssen, R. Transposable elements and the epigenetic regulation of the genome. *Nat. Rev. Genet.* **2007**, *8*, 272–285. [CrossRef] [PubMed]
44. Su, J.; Shao, X.; Liu, H.; Liu, S.; Wu, Q.; Zhang, Y. Genome-wide dynamic changes of DNA methylation of repetitive elements in human embryonic stem cells and fetal fibroblasts. *Genomics* **2012**, *99*, 10–17. [CrossRef] [PubMed]
45. Smallwood, S.A.; Kelsey, G. De novo DNA methylation: A germ cell perspective. *Trends Genet.* **2012**, *28*, 33–42. [CrossRef] [PubMed]
46. Deaton, A.M.; Bird, A. CpG islands and the regulation of transcription. *Genes Dev.* **2011**, *25*, 1010–1022. [CrossRef] [PubMed]
47. Razin, A.; Riggs, A.D. DNA methylation and gene function. *Science* **1980**, *210*, 604–610. [CrossRef] [PubMed]
48. Yang, X.; Han, H.; De Carvalho, D.D.; Lay, F.D.; Jones, P.A.; Liang, G. Gene body methylation can alter gene expression and is a therapeutic target in cancer. *Cancer Cell* **2014**, *26*, 577–590. [CrossRef] [PubMed]
49. Tate, P.H.; Bird, A.P. Effects of DNA methylation on DNA-binding proteins and gene expression. *Curr. Opin. Genet. Dev.* **1993**, *3*, 226–231. [CrossRef]
50. Yin, Y.; Morgunova, E.; Jolma, A.; Kaasinen, E.; Sahu, B.; Khund-Sayeed, S.; Das, P.K.; Kivioja, T.; Dave, K.; Zhong, F.; et al. Impact of cytosine methylation on DNA binding specificities of human transcription factors. *Science* **2017**, *356*. [CrossRef] [PubMed]
51. Schlesinger, Y.; Straussman, R.; Keshet, I.; Farkash, S.; Hecht, M.; Zimmerman, J.; Eden, E.; Yakhini, Z.; Ben-Shushan, E.; Reubinoff, B.E.; et al. Polycomb-mediated methylation on Lys27 of histone H3 pre-marks genes for de novo methylation in cancer. *Nat. Genet.* **2007**, *39*, 232–236. [CrossRef] [PubMed]
52. Lister, R.; Pelizzola, M.; Dowen, R.H.; Hawkins, R.D.; Hon, G.; Tonti-Filippini, J.; Nery, J.R.; Lee, L.; Ye, Z.; Ngo, Q.M.; et al. Human DNA methylomes at base resolution show widespread epigenomic differences. *Nature* **2009**, *462*, 315–322. [CrossRef] [PubMed]
53. Neri, F.; Rapelli, S.; Krepelova, A.; Incarnato, D.; Parlato, C.; Basile, G.; Maldotti, M.; Anselmi, F.; Oliviero, S. Intragenic DNA methylation prevents spurious transcription initiation. *Nature* **2017**, *543*, 72–77. [CrossRef] [PubMed]
54. Zhu, H.; Wang, G.H.; Qian, J. Transcription factors as readers and effectors of DNA methylation. *Nat. Rev. Genet.* **2016**, *17*, 551–565. [CrossRef] [PubMed]
55. Maunakea, A.K.; Chepelev, I.; Cui, K.; Zhao, K. Intragenic DNA methylation modulates alternative splicing by recruiting MeCP2 to promote exon recognition. *Cell. Res.* **2013**, *23*, 1256–1269. [CrossRef] [PubMed]
56. Lev Maor, G.; Yearim, A.; Ast, G. The alternative role of DNA methylation in splicing regulation. *Trends Genet.* **2015**, *31*, 274–280. [CrossRef] [PubMed]
57. Liang, G.; Weisenberger, D.J. DNA methylation aberrancies as a guide for surveillance and treatment of human cancers. *Epigenetics* **2017**, *12*, 416–432. [CrossRef] [PubMed]
58. Wagner, E.J.; Carpenter, P.B. Understanding the language of Lys36 methylation at histone H3. *Nat. Rev. Mol. Cell. Biol.* **2012**, *13*, 115–126. [CrossRef] [PubMed]
59. Ehrlich, M. DNA hypomethylation in cancer cells. *Epigenomics* **2009**, *1*, 239–259. [CrossRef] [PubMed]
60. Suzuki, H.; Watkins, D.N.; Jair, K.W.; Schuebel, K.E.; Markowitz, S.D.; Chen, W.D.; Pretlow, T.P.; Yang, B.; Akiyama, Y.; Van Engeland, M.; et al. Epigenetic inactivation of *SFRP* genes allows constitutive WNT signaling in colorectal cancer. *Nat. Genet.* **2004**, *36*, 417–422. [CrossRef] [PubMed]
61. Chen, W.Y.; Zeng, X.; Carter, M.G.; Morrell, C.N.; Chiu Yen, R.W.; Esteller, M.; Watkins, D.N.; Herman, J.G.; Mankowski, J.L.; Baylin, S.B. Heterozygous disruption of *Hic1* predisposes mice to a gender-dependent spectrum of malignant tumors. *Nat. Genet.* **2003**, *33*, 197–202. [CrossRef] [PubMed]
62. Zhou, D.; Robertson, K.D. Role of DNA methylation in genome stability. In *Genome Stability: From Virus to Human Application*, 1st ed.; Kovalchuk, I., Kovalchuk, O., Eds.; Elsevier Inc.: Woodbury, MN, USA, 2016; Volume 1, pp. 409–424.

63. Wu, X.; Zhang, Y. TET-mediated active DNA demethylation: Mechanism, function and beyond. *Nat. Rev. Genet.* **2017**, *18*, 517–534. [CrossRef] [PubMed]
64. Williams, K.; Christensen, J.; Helin, K. DNA methylation: TET proteins-guardians of CpG islands? *EMBO Rep.* **2011**, *13*, 28–35. [CrossRef] [PubMed]
65. Rasmussen, K.D.; Helin, K. Role of TET enzymes in DNA methylation, development, and cancer. *Genes Dev.* **2016**, *30*, 733–750. [CrossRef] [PubMed]
66. Pfeifer, G.P.; Kadam, S.; Jin, S.G. 5-hydroxymethylcytosine and its potential roles in development and cancer. *Epigenetics Chromatin.* **2013**, *6*, 10. [CrossRef] [PubMed]
67. Lyko, F. The DNA methyltransferase family: A versatile toolkit for epigenetic regulation. *Nat. Rev. Genet.* **2018**, *19*, 81–92. [CrossRef] [PubMed]
68. Gowher, H.; Jeltsch, A. Mammalian DNA methyltransferases: new discoveries and open questions. *Biochem Soc Trans.* **2018**. [CrossRef] [PubMed]
69. Cheng, X.; Blumenthal, R.M. Mammalian DNA methyltransferases: A structural perspective. *Structure* **2008**, *16*, 341–350. [CrossRef] [PubMed]
70. Song, J.; Rechkoblit, O.; Bestor, T.H.; Patel, D.J. Structure of DNMT1-DNA complex reveals a role for autoinhibition in maintenance DNA methylation. *Science* **2011**, *331*, 1036–1040. [CrossRef] [PubMed]
71. Li, T.; Wang, L.; Du, Y.; Xie, S.; Yang, X.; Lian, F.; Zhou, Z.; Qian, C. Structural and mechanistic insights into UHRF1-mediated DNMT1 activation in the maintenance DNA methylation. *Nucleic Acids Res.* **2018**, *46*, 3218–3231. [CrossRef] [PubMed]
72. Goll, M.G.; Kirpekar, F.; Maggert, K.A.; Yoder, J.A.; Hsieh, C.L.; Zhang, X.; Golic, K.G.; Jacobsen, S.E.; Bestor, T.H. Methylation of tRNA^{Asp} by the DNA methyltransferase homolog Dnmt2. *Science* **2006**, *311*, 395–398. [CrossRef] [PubMed]
73. Robertson, K.D.; Keyomarsi, K.; Gonzales, F.A.; Velicescu, M.; Jones, P.A. Differential mRNA expression of the human DNA methyltransferases (DNMTs) 1, 3a and 3b during the G_0/G_1 to S phase transition in normal and tumor cells. *Nucleic Acids Res.* **2000**, *28*, 2108–2113. [CrossRef] [PubMed]
74. Xie, S.; Wang, Z.; Okano, M.; Nogami, M.; Li, Y.; He, W.W.; Okumura, K.; Li, E. Cloning, expression and chromosome locations of the human *DNMT3* gene family. *Gene* **1999**, *236*, 87–95. [CrossRef]
75. Fatemi, M.; Hermann, A.; Gowher, H.; Jeltsch, A. Dnmt3a and Dnmt1 functionally cooperate during de novo methylation of DNA. *Eur. J. Biochem.* **2002**, *269*, 4981–4984. [CrossRef] [PubMed]
76. Liang, G.; Chan, M.F.; Tomigahara, Y.; Tsai, Y.C.; Gonzales, F.A.; Li, E.; Laird, P.W.; Jones, P.A. Cooperativity between DNA methyltransferases in the maintenance methylation of repetitive elements. *Mol. Cell. Biol.* **2002**, *22*, 480–491. [CrossRef] [PubMed]
77. Chen, T.; Ueda, Y.; Dodge, J.E.; Wang, Z.; Li, E. Establishment and maintenance of genomic methylation patterns in mouse embryonic stem cells by Dnmt3a and Dnmt3b. *Mol. Cell. Biol.* **2003**, *23*, 5594–5605. [CrossRef] [PubMed]
78. Baubec, T.; Colombo, D.F.; Wirbelauer, C.; Schmidt, J.; Burger, L.; Krebs, A.R.; Akalin, A.; Schubeler, D. Genomic profiling of DNA methyltransferases reveals a role for DNMT3B in genic methylation. *Nature* **2015**, *520*, 243–247. [CrossRef] [PubMed]
79. Jeong, S.; Liang, G.; Sharma, S.; Lin, J.C.; Choi, S.H.; Han, H.; Yoo, C.B.; Egger, G.; Yang, A.S.; Jones, P.A. Selective anchoring of DNA methyltransferases 3A and 3B to nucleosomes containing methylated DNA. *Mol. Cell. Biol.* **2009**, *29*, 5366–5376. [CrossRef] [PubMed]
80. Okano, M.; Bell, D.W.; Haber, D.A.; Li, E. DNA methyltransferases Dnmt3a and Dnmt3b are essential for de novo methylation and mammalian development. *Cell* **1999**, *99*, 247–257. [CrossRef]
81. Xie, Z.H.; Huang, Y.N.; Chen, Z.X.; Riggs, A.D.; Ding, J.P.; Gowher, H.; Jeltsch, A.; Sasaki, H.; Hata, K.; Xu, G.L. Mutations in DNA methyltransferase DNMT3B in ICF syndrome affect its regulation by DNMT3L. *Hum. Mol. Genet.* **2006**, *15*, 1375–1385. [CrossRef] [PubMed]
82. Ueda, Y.; Okano, M.; Williams, C.; Chen, T.; Georgopoulos, K.; Li, E. Roles for Dnmt3b in mammalian development: A mouse model for the ICF syndrome. *Development* **2006**, *133*, 1183–1192. [CrossRef] [PubMed]
83. Gagliardi, M.; Strazzullo, M.; Matarazzo, M.R. DNMT3B functions: Novel insights from human disease. *Front. Cell. Dev. Biol.* **2018**, *6*, 140. [CrossRef] [PubMed]
84. Chen, B.F.; Chan, W.Y. The de novo DNA methyltransferase DNMT3A in development and cancer. *Epigenetics* **2014**, *9*, 669–677. [CrossRef] [PubMed]

85. Challen, G.A.; Sun, D.; Mayle, A.; Jeong, M.; Luo, M.; Rodriguez, B.; Mallaney, C.; Celik, H.; Yang, L.; Xia, Z.; et al. Dnmt3a and Dnmt3b have overlapping and distinct functions in hematopoietic stem cells. *Cell. Stem Cell.* **2014**, *15*, 350–364. [CrossRef] [PubMed]
86. Xu, T.; Wang, C.; Shen, J.; Tong, P.; O'Keefe, R. Ablation of Dnmt3b in chondrocytes suppresses cell maturation during embryonic development. *J. Cell. Biochem.* **2018**. [CrossRef] [PubMed]
87. Nishikawa, K.; Iwamoto, Y.; Kobayashi, Y.; Katsuoka, F.; Kawaguchi, S.; Tsujita, T.; Nakamura, T.; Kato, S.; Yamamoto, M.; Takayanagi, H.; et al. DNA methyltransferase 3a regulates osteoclast differentiation by coupling to an S-adenosylmethionine-producing metabolic pathway. *Nat. Med.* **2015**, *21*, 281–287. [CrossRef] [PubMed]
88. Wang, C.; Abu-Amer, Y.; O'Keefe, R.J.; Shen, J. Loss of Dnmt3b in chondrocytes leads to delayed endochondral ossification and fracture repair. *J. Bone Miner. Res.* **2018**, *33*, 283–297. [CrossRef] [PubMed]
89. Hansen, R.S.; Wijmenga, C.; Luo, P.; Stanek, A.M.; Canfield, T.K.; Weemaes, C.M.; Gartler, S.M. The *DNMT3B* DNA methyltransferase gene is mutated in the ICF immunodeficiency syndrome. *Proc. Natl. Acad. Sci. USA* **1999**, *96*, 14412–14417. [CrossRef] [PubMed]
90. Song, J.; Teplova, M.; Ishibe-Murakami, S.; Patel, D.J. Structure-based mechanistic insights into DNMT1-mediated maintenance DNA methylation. *Science* **2012**, *335*, 709–712. [CrossRef] [PubMed]
91. Zhang, Z.M.; Liu, S.; Lin, K.; Luo, Y.; Perry, J.J.; Wang, Y.; Song, J. Crystal Structure of Human DNA Methyltransferase 1. *J. Mol. Biol.* **2015**, *427*, 2520–2531. [CrossRef] [PubMed]
92. Edwards, J.R.; Yarychkivska, O.; Boulard, M.; Bestor, T.H. DNA methylation and DNA methyltransferases. *Epigenetics Chromatin.* **2017**, *10*, 23. [CrossRef] [PubMed]
93. Bostick, M.; Kim, J.K.; Esteve, P.O.; Clark, A.; Pradhan, S.; Jacobsen, S.E. UHRF1 plays a role in maintaining DNA methylation in mammalian cells. *Science* **2007**, *317*, 1760–1764. [CrossRef] [PubMed]
94. Chuang, L.S.; Ian, H.I.; Koh, T.W.; Ng, H.H.; Xu, G.; Li, B.F. Human DNA-(cytosine-5) methyltransferase-PCNA complex as a target for p21^{WAF1}. *Science* **1997**, *277*, 1996–2000. [CrossRef] [PubMed]
95. Schermelleh, L.; Haemmer, A.; Spada, F.; Rosing, N.; Meilinger, D.; Rothbauer, U.; Cardoso, M.C.; Leonhardt, H. Dynamics of Dnmt1 interaction with the replication machinery and its role in postreplicative maintenance of DNA methylation. *Nucleic Acids Res.* **2007**, *35*, 4301–4312. [CrossRef] [PubMed]
96. Gaudet, F.; Hodgson, J.G.; Eden, A.; Jackson-Grusby, L.; Dausman, J.; Gray, J.W.; Leonhardt, H.; Jaenisch, R. Induction of tumors in mice by genomic hypomethylation. *Science* **2003**, *300*, 489–492. [CrossRef] [PubMed]
97. Li, E.; Zhang, Y. DNA methylation in mammals. *Cold Spring Harb Perspect Biol* **2014**, *6*, a019133. [CrossRef] [PubMed]
98. Sharif, J.; Muto, M.; Takebayashi, S.; Suetake, I.; Iwamatsu, A.; Endo, T.A.; Shinga, J.; Mizutani-Koseki, Y.; Toyoda, T.; Okamura, K.; et al. The SRA protein Np95 mediates epigenetic inheritance by recruiting Dnmt1 to methylated DNA. *Nature* **2007**, *450*, 908–912. [CrossRef] [PubMed]
99. Chen, T.; Hevi, S.; Gay, F.; Tsujimoto, N.; He, T.; Zhang, B.; Ueda, Y.; Li, E. Complete inactivation of DNMT1 leads to mitotic catastrophe in human cancer cells. *Nat. Genet.* **2007**, *39*, 391–396. [CrossRef] [PubMed]
100. Zhang, W.; Xu, J. DNA methyltransferases and their roles in tumorigenesis. *Biomark Res.* **2017**, *5*. [CrossRef] [PubMed]
101. Mudbhary, R.; Hoshida, Y.; Chernyavskaya, Y.; Jacob, V.; Villanueva, A.; Fiel, M.I.; Chen, X.; Kojima, K.; Thung, S.; Bronson, R.T.; et al. UHRF1 overexpression drives DNA hypomethylation and hepatocellular carcinoma. *Cancer Cell* **2014**, *25*, 196–209. [CrossRef] [PubMed]
102. Jia, Y.; Li, P.; Fang, L.; Zhu, H.; Xu, L.; Cheng, H.; Zhang, J.; Li, F.; Feng, Y.; Li, Y.; et al. Negative regulation of DNMT3A de novo DNA methylation by frequently overexpressed UHRF family proteins as a mechanism for widespread DNA hypomethylation in cancer. *Cell Discov.* **2016**, *2*, 16007. [CrossRef] [PubMed]
103. Ko, Y.G.; Nishino, K.; Hattori, N.; Arai, Y.; Tanaka, S.; Shiota, K. Stage-by-stage change in DNA methylation status of Dnmt1 locus during mouse early development. *J. Biol. Chem.* **2005**, *280*, 9627–9634. [CrossRef] [PubMed]
104. Howell, C.Y.; Bestor, T.H.; Ding, F.; Latham, K.E.; Mertineit, C.; Trasler, J.M.; Chaillet, J.R. Genomic imprinting disrupted by a maternal effect mutation in the *Dnmt1* gene. *Cell* **2001**, *104*, 829–838. [CrossRef]
105. Suetake, I.; Mishima, Y.; Kimura, H.; Lee, Y.H.; Goto, Y.; Takeshima, H.; Ikegami, T.; Tajima, S. Characterization of DNA-binding activity in the N-terminal domain of the DNA methyltransferase Dnmt3a. *Biochem. J.* **2011**, *437*, 141–148. [CrossRef] [PubMed]

106. Chen, T.; Ueda, Y.; Xie, S.; Li, E. A novel Dnmt3a isoform produced from an alternative promoter localizes to euchromatin and its expression correlates with active de novo methylation. *J. Biol. Chem.* **2002**, *277*, 38746–38754. [CrossRef] [PubMed]
107. Nguyen, S.; Meletis, K.; Fu, D.; Jhaveri, S.; Jaenisch, R. Ablation of de novo DNA methyltransferase Dnmt3a in the nervous system leads to neuromuscular defects and shortened lifespan. *Dev. Dyn.* **2007**, *236*, 1663–1676. [CrossRef] [PubMed]
108. Tatton-Brown, K.; Seal, S.; Ruark, E.; Harmer, J.; Ramsay, E.; Duarte, S.D.; Zachariou, A.; Hanks, S.; O'Brien, E.; Aksglaede, L.; et al. Mutations in the DNA methyltransferase gene *DNMT3A* cause an overgrowth syndrome with intellectual disability. *Nat. Genet.* **2014**, *46*, 385–388. [CrossRef] [PubMed]
109. Russler-Germain, D.A.; Spencer, D.H.; Young, M.A.; Lamprecht, T.L.; Miller, C.A.; Fulton, R.; Meyer, M.R.; Erdmann-Gilmore, P.; Townsend, R.R.; Wilson, R.K.; et al. The R882H DNMT3A mutation associated with AML dominantly inhibits wild-type DNMT3A by blocking its ability to form active tetramers. *Cancer Cell* **2014**, *25*, 442–454. [CrossRef] [PubMed]
110. Ostler, K.R.; Davis, E.M.; Payne, S.L.; Gosalia, B.B.; Exposito-Cespedes, J.; Le Beau, M.M.; Godley, L.A. Cancer cells express aberrant DNMT3B transcripts encoding truncated proteins. *Oncogene* **2007**, *26*, 5553–5563. [CrossRef] [PubMed]
111. Wang, L.; Wang, J.; Sun, S.; Rodriguez, M.; Yue, P.; Jang, S.J.; Mao, L. A novel DNMT3B subfamily, ΔDNMT3B, is the predominant form of DNMT3B in non-small cell lung cancer. *Int. J. Oncol.* **2006**, *29*, 201–207. [CrossRef] [PubMed]
112. Gopalakrishnan, S.; Van Emburgh, B.O.; Shan, J.; Su, Z.; Fields, C.R.; Vieweg, J.; Hamazaki, T.; Schwartz, P.H.; Terada, N.; Robertson, K.D. A novel DNMT3B splice variant expressed in tumor and pluripotent cells modulates genomic DNA methylation patterns and displays altered DNA binding. *Mol. Cancer Res.* **2009**, *7*, 1622–1634. [CrossRef] [PubMed]
113. Duymich, C.E.; Charlet, J.; Yang, X.; Jones, P.A.; Liang, G. DNMT3B isoforms without catalytic activity stimulate gene body methylation as accessory proteins in somatic cells. *Nat. Commun.* **2016**, *7*, 11453. [CrossRef] [PubMed]
114. Barau, J.; Teissandier, A.; Zamudio, N.; Roy, S.; Nalesso, V.; Herault, Y.; Guillou, F.; Bourc'his, D. The DNA methyltransferase DNMT3C protects male germ cells from transposon activity. *Science* **2016**, *354*, 909–912. [CrossRef] [PubMed]
115. Bourc'his, D.; Xu, G.L.; Lin, C.S.; Bollman, B.; Bestor, T.H. Dnmt3L and the establishment of maternal genomic imprints. *Science* **2001**, *294*, 2536–2539. [CrossRef] [PubMed]
116. Chedin, F. The DNMT3 family of mammalian de novo DNA methyltransferases. *Prog. Mol. Biol. Transl. Sci.* **2011**, *101*, 255–285. [CrossRef] [PubMed]
117. Bourc'his, D.; Bestor, T.H. Meiotic catastrophe and retrotransposon reactivation in male germ cells lacking Dnmt3L. *Nature* **2004**, *431*, 96–99. [CrossRef] [PubMed]
118. La Salle, S.; Mertineit, C.; Taketo, T.; Moens, P.B.; Bestor, T.H.; Trasler, J.M. Windows for sex-specific methylation marked by DNA methyltransferase expression profiles in mouse germ cells. *Dev. Biol.* **2004**, *268*, 403–415. [CrossRef] [PubMed]
119. Hata, K.; Okano, M.; Lei, H.; Li, E. Dnmt3L cooperates with the Dnmt3 family of de novo DNA methyltransferases to establish maternal imprints in mice. *Development* **2002**, *129*, 1983–1993. [PubMed]
120. Hu, Y.G.; Hirasawa, R.; Hu, J.L.; Hata, K.; Li, C.L.; Jin, Y.; Chen, T.; Li, E.; Rigolet, M.; Viegas-Pequignot, E.; et al. Regulation of DNA methylation activity through Dnmt3L promoter methylation by Dnmt3 enzymes in embryonic development. *Hum. Mol. Genet.* **2008**, *17*, 2654–2664. [CrossRef] [PubMed]
121. Ooi, S.K.; Qiu, C.; Bernstein, E.; Li, K.; Jia, D.; Yang, Z.; Erdjument-Bromage, H.; Tempst, P.; Lin, S.P.; Allis, C.D.; et al. DNMT3L connects unmethylated lysine 4 of histone H3 to de novo methylation of DNA. *Nature* **2007**, *448*, 714–717. [CrossRef] [PubMed]
122. Veland, N.; Lu, Y.; Hardikar, S.; Gaddis, S.; Zeng, Y.; Liu, B.; Estecio, M.R.; Takata, Y.; Lin, K.; Tomida, M.W.; et al. DNMT3L facilitates DNA methylation partly by maintaining DNMT3A stability in mouse embryonic stem cells. *Nucleic Acids Res.* **2018**. [CrossRef] [PubMed]
123. Neri, F.; Krepelova, A.; Incarnato, D.; Maldotti, M.; Parlato, C.; Galvagni, F.; Matarese, F.; Stunnenberg, H.G.; Oliviero, S. Dnmt3L antagonizes DNA methylation at bivalent promoters and favors DNA methylation at gene bodies in ESCs. *Cell* **2013**, *155*, 121–134. [CrossRef] [PubMed]

124. Minami, K.; Chano, T.; Kawakami, T.; Ushida, H.; Kushima, R.; Okabe, H.; Okada, Y.; Okamoto, K. DNMT3L is a novel marker and is essential for the growth of human embryonal carcinoma. *Clin. Cancer Res.* **2010**, *16*, 2751–2759. [CrossRef] [PubMed]
125. Chedin, F.; Lieber, M.R.; Hsieh, C.L. The DNA methyltransferase-like protein DNMT3L stimulates de novo methylation by Dnmt3a. *Proc. Natl. Acad. Sci. USA* **2002**, *99*, 16916–16921. [CrossRef] [PubMed]
126. Gordon, C.A.; Hartono, S.R.; Chedin, F. Inactive DNMT3B splice variants modulate de novo DNA methylation. *PLoS ONE* **2013**, *8*, e69486. [CrossRef] [PubMed]
127. Tiedemann, R.L.; Hlady, R.A.; Hanavan, P.D.; Lake, D.F.; Tibes, R.; Lee, J.H.; Choi, J.H.; Ho, T.H.; Robertson, K.D. Dynamic reprogramming of DNA methylation in SETD2-deregulated renal cell carcinoma. *Oncotarget* **2016**, *7*, 1927–1946. [CrossRef] [PubMed]
128. Sharma, S.; De Carvalho, D.D.; Jeong, S.; Jones, P.A.; Liang, G. Nucleosomes containing methylated DNA stabilize DNA methyltransferases 3A/3B and ensure faithful epigenetic inheritance. *PLoS Genet.* **2011**, *7*, e1001286. [CrossRef] [PubMed]
129. Chen, Z.X.; Mann, J.R.; Hsieh, C.L.; Riggs, A.D.; Chedin, F. Physical and functional interactions between the human DNMT3L protein and members of the de novo methyltransferase family. *J. Cell. Biochem.* **2005**, *95*, 902–917. [CrossRef] [PubMed]
130. Okano, M.; Xie, S.; Li, E. Cloning and characterization of a family of novel mammalian DNA (cytosine-5) methyltransferases. *Nat. Genet.* **1998**, *19*, 219–220. [CrossRef] [PubMed]
131. Choi, S.H.; Heo, K.; Byun, H.M.; An, W.; Lu, W.; Yang, A.S. Identification of preferential target sites for human DNA methyltransferases. *Nucleic Acids Res.* **2011**, *39*, 104–118. [CrossRef] [PubMed]
132. Weisenberger, D.J.; Velicescu, M.; Cheng, J.C.; Gonzales, F.A.; Liang, G.; Jones, P.A. Role of the DNA methyltransferase variant DNMT3b3 in DNA methylation. *Mol. Cancer Res.* **2004**, *2*, 62–72. [PubMed]
133. Van Emburgh, B.O.; Robertson, K.D. Modulation of Dnmt3b function in vitro by interactions with Dnmt3L, Dnmt3a and Dnmt3b splice variants. *Nucleic Acids Res.* **2011**, *39*, 4984–5002. [CrossRef] [PubMed]
134. Saito, Y.; Kanai, Y.; Sakamoto, M.; Saito, H.; Ishii, H.; Hirohashi, S. Overexpression of a splice variant of DNA methyltransferase 3b, DNMT3b4, associated with DNA hypomethylation on pericentromeric satellite regions during human hepatocarcinogenesis. *Proc. Natl. Acad. Sci. USA* **2002**, *99*, 10060–10065. [CrossRef] [PubMed]
135. Ostler, K.R.; Yang, Q.; Looney, T.J.; Zhang, L.; Vasanthakumar, A.; Tian, Y.; Kocherginsky, M.; Raimondi, S.L.; DeMaio, J.G.; Salwen, H.R.; et al. Truncated DNMT3B isoform DNMT3B7 suppresses growth, induces differentiation, and alters DNA methylation in human neuroblastoma. *Cancer Res.* **2012**, *72*, 4714–4723. [CrossRef] [PubMed]
136. Wang, J.; Walsh, G.; Liu, D.D.; Lee, J.J.; Mao, L. Expression of ΔDNMT3B variants and its association with promoter methylation of *p16* and *RASSF1A* in primary non-small cell lung cancer. *Cancer Res.* **2006**, *66*, 8361–8366. [CrossRef] [PubMed]
137. Wang, J.; Bhutani, M.; Pathak, A.K.; Lang, W.; Ren, H.; Jelinek, J.; He, R.; Shen, L.; Issa, J.P.; Mao, L. ΔDNMT3B variants regulate DNA methylation in a promoter-specific manner. *Cancer Res.* **2007**, *67*, 10647–10652. [CrossRef] [PubMed]
138. Qiu, C.; Sawada, K.; Zhang, X.; Cheng, X. The PWWP domain of mammalian DNA methyltransferase Dnmt3b defines a new family of DNA-binding folds. *Nat. Struct. Biol.* **2002**, *9*, 217–224. [CrossRef] [PubMed]
139. Qin, S.; Min, J. Structure and function of the nucleosome-binding PWWP domain. *Trends Biochem. Sci.* **2014**, *39*, 536–547. [CrossRef] [PubMed]
140. Li, Z.; Dai, H.; Martos, S.N.; Xu, B.; Gao, Y.; Li, T.; Zhu, G.; Schones, D.E.; Wang, Z. Distinct roles of DNMT1-dependent and DNMT1-independent methylation patterns in the genome of mouse embryonic stem cells. *Genome Biol.* **2015**, *16*, 115. [CrossRef] [PubMed]
141. Elliott, E.N.; Sheaffer, K.L.; Kaestner, K.H. The 'de novo' DNA methyltransferase Dnmt3b compensates the Dnmt1-deficient intestinal epithelium. *Elife* **2016**, *5*. [CrossRef] [PubMed]
142. Jones, P.A.; Taylor, S.M. Cellular-differentiation, cytidine analogs and DNA methylation. *Cell* **1980**, *20*, 85–93. [CrossRef]
143. Kaminskas, E.; Farrell, A.T.; Wang, Y.C.; Sridhara, R.; Pazdur, R. FDA drug approval summary: Azacitidine (5-azacytidine, Vidaza) for injectable suspension. *Oncologist* **2005**, *10*, 176–182. [CrossRef] [PubMed]

144. Pinto, A.; Zagonel, V. 5-Aza-2′-deoxycytidine (Decitabine) and 5-azacytidine in the treatment of acute myeloid leukemias and myelodysplastic syndromes: past, present and future trends. *Leukemia* **1993**, *7*, 51–60. [PubMed]
145. Atallah, E.; Kantarjian, H.; Garcia-Manero, G. The role of decitabine in the treatment of myelodysplastic syndromes. *Expert Opin. Pharmacother.* **2007**, *8*, 65–73. [CrossRef] [PubMed]
146. Jones, P.A.; Issa, J.P.; Baylin, S. Targeting the cancer epigenome for therapy. *Nat. Rev. Genet.* **2016**, *17*, 630–641. [CrossRef] [PubMed]
147. Thinnes, C.C.; England, K.S.; Kawamura, A.; Chowdhury, R.; Schofield, C.J.; Hopkinson, R.J. Targeting histone lysine demethylases—Progress, challenges, and the future. *Bba-Gene Regul. Mech.* **2014**, *1839*, 1416–1432. [CrossRef] [PubMed]
148. Lei, Y.; Zhang, X.; Su, J.; Jeong, M.; Gundry, M.C.; Huang, Y.H.; Zhou, Y.; Li, W.; Goodell, M.A. Targeted DNA methylation in vivo using an engineered dCas9-MQ1 fusion protein. *Nat. Commun.* **2017**, *8*, 16026. [CrossRef] [PubMed]
149. Huang, Y.H.; Su, J.Z.; Lei, Y.; Brunetti, L.; Gundry, M.C.; Zhang, X.T.; Jeong, M.; Li, W.; Goodell, M.A. DNA epigenome editing using CRISPR-Cas SunTag-directed DNMT3A. *Genome Biol.* **2017**, *18*. [CrossRef] [PubMed]
150. Xu, X.; Tao, Y.; Gao, X.; Zhang, L.; Li, X.; Zou, W.; Ruan, K.; Wang, F.; Xu, G.L.; Hu, R. A CRISPR-based approach for targeted DNA demethylation. *Cell. Discov.* **2016**, *2*, 16009. [CrossRef] [PubMed]

© 2019 by the authors. Licensee MDPI, Basel, Switzerland. This article is an open access article distributed under the terms and conditions of the Creative Commons Attribution (CC BY) license (http://creativecommons.org/licenses/by/4.0/).

Review

Effect of Disease-Associated Germline Mutations on Structure Function Relationship of DNA Methyltransferases

Allison B. Norvil, Debapriya Saha, Mohd Saleem Dar and Humaira Gowher *

Department of Biochemistry, Purdue University, West Lafayette, IN 47907, USA; anorvil@purdue.edu (A.B.N.); saha27@purdue.edu (D.S.); darm@purdue.edu (M.S.D.)
* Correspondence: hgowher@purdue.edu

Received: 27 March 2019; Accepted: 8 May 2019; Published: 14 May 2019

Abstract: Despite a large body of evidence supporting the role of aberrant DNA methylation in etiology of several human diseases, the fundamental mechanisms that regulate the activity of mammalian DNA methyltransferases (DNMTs) are not fully understood. Recent advances in whole genome association studies have helped identify mutations and genetic alterations of DNMTs in various diseases that have a potential to affect the biological function and activity of these enzymes. Several of these mutations are germline-transmitted and associated with a number of hereditary disorders, which are potentially caused by aberrant DNA methylation patterns in the regulatory compartments of the genome. These hereditary disorders usually cause neurological dysfunction, growth defects, and inherited cancers. Biochemical and biological characterization of DNMT variants can reveal the molecular mechanism of these enzymes and give insights on their specific functions. In this review, we introduce roles and regulation of DNA methylation and DNMTs. We discuss DNMT mutations that are associated with rare diseases, the characterized effects of these mutations on enzyme activity and provide insights on their potential effects based on the known crystal structure of these proteins.

Keywords: ADCA-DN; HSAN1E; TBRS; dwarfism; DNMT3A; DNMT1; rare diseases; PCC/PGL; DNA methylation

1. Introduction

DNA methylation is a highly conserved epigenetic modification in mammals and takes place at the 5′ position of cytosine, largely at the CpG dinucleotide [1,2]. The distribution of DNA methylation in mammalian genomes is bimodal such that the repetitive elements and transposons are most densely methylated and the regions with highest propensity of CpG (CpG islands) are least methylated [3]. DNA methylation increases the information content of the genome through its potential to control gene expression. In regulatory elements of genes, including promoters and enhancers, DNA methylation is largely associated with repressed genes and is often tissue specific. Conversely, high DNA methylation is found in gene bodies of highly transcribing genes. All these observations indicate that interpretation of DNA methylation is dependent on the genomic context. Despite the complexity of DNA demethylation, the active loss of DNA methylation has been observed both during early development and at certain inducible genes in later adulthood [4]. Although a functional demethylase that can directly remove -CH3 groups from the 5′C of cytosine has not been discovered, reversal of DNA methylation can be mediated by the conversion of methyl to hydoxymethyl and higher oxidation states by the Tet family of methylcytosine dioxygenases. This can lead to progressive loss of the modification because these oxidized states unlike DNA methylation cannot be maintained. Besides being an intermediate of DNA demethylation process, hydroxymethylation at regulatory elements, which are in the *primed state*, alter the signal output of DNA methylation by changing its detectability [5–8].

DNA methyltransferases (DNMTs) are a class of enzymes that catalyze the transfer of a methyl group from S-Adenosyl-L-methionine (AdoMet) to DNA. Mammalian DNMTs belong to two structurally and functionally distinct families, DNMT1 and DNMT3 [9–11]. In somatic cells, 60–80% of CpG sites are methylated and DNMT1 diligently copies the methylation patterns from the parent to the daughter strand post-replication and repair [1,12,13]. This activity particularly ensures the maintenance of unmethylated regions by averting spurious de novo DNA methylation. The DNMT3 family includes DNMT3A and DNMT3B, which are the de novo methyltransferases, and one regulatory factor, DNMT3-Like protein (DNMT3L) [11]. DNMT3L is catalytically inactive but interacts with both DNMT3A and 3B to enhance their enzymatic activity [14]. The DNMT3 proteins are required for the establishment of genomic DNA methylation during embryogenesis after its erasure at the preimplantation stage [13]. Whereas DNMT1 is ubiquitously expressed, DNMT3 enzymes show distinct tissue specific expression and methylate the regulatory elements of the transcriptionally inactive genes. Several lines of evidence including distinct phenotype of DNMT3A and DNMT3B KO mice indicate distinct biological functions for these enzymes [15]. Studies of molecular and cellular phenotypes resulting from DNMT mutations have largely contributed to our understanding of the biological roles of DNA methylation [16].

Mammalian DNMTs constitute a C-terminal catalytic domain, a structure which is highly conserved from bacteria to humans [9,17]. Given mammalian DNMTs have weak sequence specificity, their target site recognition is guided by the N-terminal regulatory region, which interacts with transcription factors, chromatin binding proteins, and histone-tail modifications [18]. Consequently, these interactions regulate site-specific DNA methylation leading to differential gene expression. Abnormal patterns of DNA methylation have been observed in several diseases and in all types of cancer. This could be caused by either loss of function of DNMTs or their interactions with modulators [19,20]. Recent high throughput sequencing have revealed mutations in DNMTs associated with several diseased states. Interestingly, in all reported disorders only one of the three enzymes accumulates mutations, leading to distinct phenotypes [21]. Although these mutations are distributed throughout the *DNMT* gene, most of them tend to cluster in the functional domains of these enzymes. The adverse effect of these mutations on the catalytic activity and function of DNMTs has been established in multiple reports [22–26]. The following content will discuss recent advancements in the investigation of etiological consequences of germline–transmitted mutations in DNMT1 and DNMT3A and the effect of these mutations on catalytic and targeting mechanism of the enzymes.

2. Structural and Functional Alterations of DNMT1 by Disease Associated Mutations

The eukaryotic DNMT1 is a multimodular protein comprising of a replication foci-targeting sequence (RFTS), a DNA binding CXXC domain, two bromo-adjacent homology (BAH) domains and a C-terminal catalytic domain (Figure 1A) [27]. DNMT1 has an intrinsic preference for hemimethylated CpG sites. This preference is further modulated by interactions of both CXXC and RFTS domains with the DNA binding region of DNMT1 leading to autoinhibition. The CXXC domain binds to unmethylated CpG dinucleotides and sandwiches a section of highly acidic amino acids (the autoinhibitory BAH1-CXXC linker) between the DNA and DNMT1 active site. The autoinhibition by RFTS is relieved by its interaction with UHRF1 (ubiquitin-like, containing PHD and RING finger domains 1), which binds to hemimethylated CpG dinucleotides. It is suggested that hemimethylated DNA from UHRF1 is transferred to the active site of DNMT1 after the inhibitory RFTS has been displaced by the UHRF1/hemimethylated DNA complex [28]. The multiple functional domains of the N- terminus have different roles including coordination of methylation and replication during S-phase, partial suppression of de novo methylation and nuclear localization [2,11,29].

Exome sequencing studies revealed several mutations in DNMT1 that result in two adult onset, progressive neurological disorders. These mutations are germline dominant and include 13 amino acid substitutions in the RFTS domain that potentially disrupt the catalytic activity of the enzyme (Figure 1A). The first is hereditary sensory neuropathy with dementia and hearing loss (HSAN1E) and the second is autosomal dominant cerebellar ataxia, deafness and narcolepsy (ADCA-DN), caused by progressive loss of sensory neuron function [30–32].

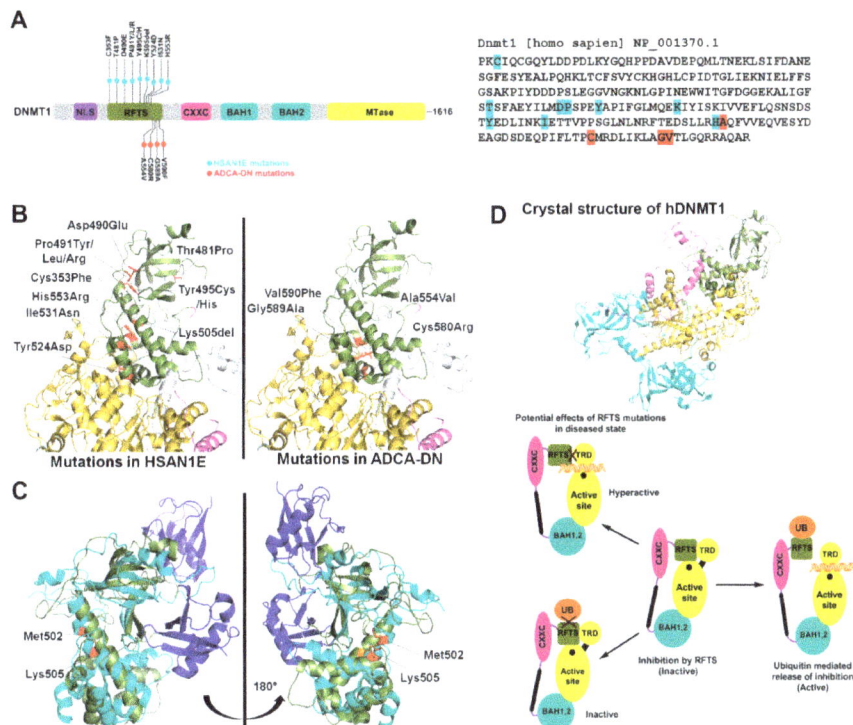

Figure 1. HSAN1E (hereditary sensory neuropathy with dementia and hearing loss) and ADCA-DN (autosomal dominant cerebellar ataxia, deafness and narcolepsy) mutations in DNMT1 (A) Left Schematic representation of the *hDNMT1* gene. HSAN1E mutations are listed above the gene in light blue, while ADCA-DN mutations are listed below the gene in red. Right Nucleotide sequence of the RFTS (replication foci-targeting sequence) domain, with the mutations highlighted in color corresponding to the schematic. (B) Crystal structure of hDNMT1 (351–1600) from the Protein Data Bank (PDB: 4WXX). The cartoon structure of the RFTS domain is green, the CXXC domain is purple, and the MTase (methyltransferase) domain is yellow. All disease mutations are located in the RFTS domain, and are shown as stick structures in red. The positions of HSAN1E and ADCA-DN mutations are shown in the left and right DNMT1 structure respectively. (C) Overlay of the hDNMT1 RFTS domain bound (light blue) and unbound (green) to two molecules of ubiquitin (dark blue) from the PDB: 4WXX and 5YDR. When ubiquitin is bound, the RFTS domain bends about 30° at Met502. The HSAN1E mutation Lys505del and Met502 are shown in red in RFTS domain bound to ubiquitin and in orange in the unbound RFTS domain. (D) Model showing the effect of mutations in the RFTS domain on the catalytic mechanism of DNMT1. DNMT1 is auto-inhibited by the interaction of its RFTS domain with the target recognition domain (TRD) in MTase domain that prevents DNA binding. When RFTS interacts with ubiquitin, auto-inhibition is released allowing TRD to interact with the hemi-methylated DNA. However, mutations in the RFTS that alter its binding to ubiquitin will prevent enzyme activation, while mutations that alter its binding to the TRD will leave the enzyme in a hyperactive state.

2.1. HSAN1E

To date, nine heterozygous mutations of the *DNMT1* gene have been identified in HSAN1E patients. These mutations are located mostly in exon 20, which encodes part of the RFTS domain (Figure 1A). The first discovered mutations include substitution of two contiguous amino acids Asp490Glu and Pro491Arg, and a Tyr495Cys substitution. Systematic investigation revealed that these mutations lead to protein degradation, reduced DNMT1 activity and defective binding to heterochromatin in G2 phase [32]. This ultimately leads to widespread DNA hypomethylation including pericentromeric satellite 2 sequences, other repetitive elements, intergenic regions, imprinted genes, and transcriptional start sites. At some CpG islands, site-specific hypermethylation was also reported [33]. The mechanism explaining hypermethylation can be interpreted from a study in cancer cells showing that the deletion of RFTS domain makes DNMT1 hyperactive and available for euchromatic binding. This was suggested to be due to loss of RFTS interaction with an unknown heterochromatin binding protein leading to aberrant localization. A point mutation that could have a similar effect on RFTS's interaction with heterochromatin could potentially cause hypermethylation at CpG islands. [34]. Studies in mouse embryonic stem cells (ESCs) revealed that DNMT1 mutations Pro491Tyr and Tyr495Cys lead to decreased binding with the E3 ubiquitin ligase UHRF1. The ESCs overexpressing the variant DNMT1 enzymes failed to properly differentiate into neuronal progenitor cells, suggesting a differentiation defect as a possible mechanism for disease progression [35]. In addition to the aforementioned mutations, six other mutations in RFTS domain were identified in HSAN1E patients (Figure 1A) [30,36–38]. However, in absence of their biochemical characterization, the role of several of these mutations in disease development is not understood. Based on crystal structure analysis, we speculate that these mutations may lead to altered domain structure or interfere with protein-protein interactions (Figure 1B) [39]. The UHRF1 interaction region of DNMT1 spans from residues 458–500, suggesting that Thr481Pro mutation may also cause decreased UHRF1 binding. Biochemical studies however show no effect of this mutation on its localization to replication foci [30]. UHRF1 interacts with DNMT1 through its N-terminal UBL (ubiquitin like) domain. Crystal structure of RFTS domain co-crystalized with ubiquitin shows interaction with two ubiquitin molecules [40]. To accommodate the binding of two ubiquitin molecules, the RFTS domain undergoes a drastic conformational change, bending an α helix by about 30° at Met502 (Figure 1C) [40]. Deletion of Lys505 may prevent the conformational change and affect the RFTS-ubiquitin interaction [40] (Figure 1C). His553, which is located in exon 21, interacts with Glu504 and Lys505 and may facilitate this conformational change, which can be affected by His553Arg mutation [38] (Figure 1C). Based on its position in RFTS domain, the Cys353Phe substitution may perturb zinc binding in the RFTS domain and/or affect protein stability (Figure 1B). Indeed, a study using recombinant expression of many of these variants showed cytosolic aggregation and early degradation of the GFP-tagged mutant proteins [30]. Given that, DNMT1 protein is present in appreciable levels in neurons, cellular toxicity caused by protein aggregates, may underlie some clinical manifestations [30]. Together these studies support the conclusion that loss of DNMT1 targeting causes site-specific changes in DNA methylation in the HSAN1E patients.

2.2. ADCA-DN

ADCA-DN patients are reported to have four missense substitutions, Ala554Val, Cys580Arg, Gly589Ala, and Val590Phe. Similar to HSAN1E, all four mutations map to the RFTS domain of DNMT1, however they occur exclusively in exon 21 [31,37,41]. All four mutations are located in the α helical bundle of the RFTS C-lobe, which has a hydrophobic pocket at the center. Therefore substitutions, Ala554Val and Val590Phe, present in the hydrophobic pocket and Gly589Ala, closely located to the hydrophobic pocket, may destabilize the RFTS domain (Figure 1B). Further, these substitutions could also impair autoinhibition by weakening the interaction of the RFTS C-lobe with the DNA binding region of the MTase domain, rendering the enzyme hyperactive with the potential to be mistargeted (Figure 1D). This speculation is supported by data showing that the truncation of RFTS domain leads to dysregulation of DNMT1 activity [34]. Methylation profiling of ADCA-DN patients showed global hypomethylation and hypermethylation specifically at around 80 CpG islands of which nearly half were associated with promoters and rest were inter- or intragenic. The differentially methylated regions were enriched in genes for cellular and anatomical developmental processes [42]. However, the effect of these changes on expression of associated genes and consequent biological function is unknown.

Besides HSAN1E and ADCA-DN disorders, previous work has established the role of aberrant DNA methylation in neurological disorders, such as Alzheimer's and Parkinson's disease [43–47]. However, the effect of aberrant DNA methylation is potentially due to mis-regulation of DNMTs and/or interactions of methyl-CpG-binding domain (MBD) proteins, such as, MeCP2 with methylated DNA [48]. Mutation of DNMT1 also cause alterations in the genome-wide DNA methylation patterns in colorectal cancer patients [49], however none of the patients with neurological disorders were shown to develop cancer [5,50]. In summary, these studies support that germline versus somatic mutations have a spatiotemporal effect on the activity of DNMT1 during development and adulthood.

3. Structural and Functional Alterations of DNMT3 by Disease Associated Mutations

The DNMT3 family consists of two catalytically active DNMTs, DNMT3A and DNMT3B and a catalytically inactive protein, DNMT3L. DNMT3A and DNMT3B have similar domain organization; both have a variable region at the N-terminus, followed by the Pro-Trp-Trp-Pro (PWWP) domain, a Cys-rich Zn-binding domain also called ATRX-DNMT3-DNMT3L (ADD) domain and a C-terminal methyltransferase (MTase) domain [51]. The PWWP domain targets DNMT3A activity by binding to DNA and histone H3 methylated at the Lys36 residue (H3K36me2/3) [52–54]. Co-crystal structure of the DNMT3L–ADD domain with histone H3 peptide shows that it specifically interacts with the Lys4 residue only when it is unmethylated (H3K4me0). Methylation of histone H3K4 (H3K4me1/2/3) disrupts this interaction [55]. Interaction of the ADD domain with the DNA binding region of the DNMT3A catalytic domain was revealed in a recent crystal structure suggesting its role in autoinhibition of the DNMT3A enzymatic activity. This autoinhibition is relieved by the interaction of the DNMT3A-ADD domain with histone H3K4me0 [56]. The dynamic role of this regulatory mechanism was shown to regulate DNA methylation at the enhancers of pluripotency genes during embryonic stem cell differentiation [57].

The MTase domain comprises ten sequence motifs, which are conserved in all cytosine DNMTs and a have direct role in catalysis [9,17]. Motifs I–III are involved in binding to the AdoMet, whereas motifs IV and VI are required for the catalysis. The region between and including motifs VIII and IX is called the target recognition domain (TRD) and is responsible for DNA binding in DNMT3A [58,59]. DNMT3A forms a hetero-tetrameric structure with DNMT3L in which two DNMT3A monomers form the center of the complex, flanked by two DNMT3L monomers on either side [60]. In the heterotetramer of the mouse protein, the DNMT3A–3L interaction is mediated by two Phe residues (261 in DNMT3L and 728 in DNMT3A) and DNMT3A–3A interaction surface comprises Arg881 and Asp872, therefore named as the RD interface [14,60]. In the absence of DNMT3L, DNMT3A forms homo-tetramers and can oligomerize on DNA [26,59]. This property facilitates DNMT3A's cooperativity, where multiple enzyme units interact with DNA to methylate it at a faster rate [61]. Mutations in the RD interface disrupts DNMT3A DNA binding and activity demonstrating the critical role of protein dimerization in catalysis [26].

While there are only a handful of diseases caused by DNMT3 germline mutations, these diseases are caused by a plethora of mutations. Among all DNMTs, the disease-causing mutations were first discovered in DNMT3B in patients with immunodeficiency, centromeric instability, and facial anomalies (ICF) syndrome [15,62,63]. The implications of these mutations on DNMT3B activity and on the etiology of ICF has been extensively investigated and reviewed [64]. More recently, a high prevalence of DNMT3A somatic mutations were observed in hematological malignancies, Acute Myeloid Leukemia (AML) and Myelodysplastic syndrome (MDS). A series of germline mutations in DNMT3A were discovered in patients with growth syndromes, Tatton-Brown-Rahman syndrome (TBRS) and microcephalic dwarfism (MD). Some of these mutations were also found in hereditary tumors, pheochromocytomas PCC) and paragangliomas (PGL).

3.1. Tatton-Brown-Rahman Syndrome

DNMT3A related overgrowth syndrome, also known as Tatton-Brown-Rahman syndrome (TBRS) is an autosomal dominant condition characterized by overgrowth, distinctive facial appearance, and intellectual disability. It is caused by heterozygous mutations in DNMT3A that are transmitted through the germ line. In 55 TBRS patients, more than 40 distinct DNMT3A variants have been reported. Of these, most are missense mutations (30 variants), and the rest are nonsense variants, frameshift variants or whole gene deletions [65–72]. The TBRS mutations are specifically localized in each of the three functional domains of DNMT3A, 11 of which overlap with the somatic DNMT3A variants found in hematological malignancies (Figure 2A). A recent study performed a genome-wide DNA methylation analysis of 16 TBRS patients and detailed analysis of the methylation distribution in one patient with Arg771Gln substitution. Their data showed widespread DNA hypomethylation at specific genomic sites located near genes involved in morphogenesis, development, differentiation, and malignancy predisposition pathways, thus providing an important insight into developmental mechanisms that are dysregulated in the disease [73].

In the PWWP domain, several frameshift mutations, one deletion, and five missense substitutions were reported (Figure 2A). Analysis of the crystal structure shows that the missense mutations cluster around the aromatic cage which interacts with H3K36me2/3 [54]. Notable mutations include Arg301Trp, Gly298Trp/Arg, Tyr365Cys and Trp297del, that are near to or interact with the aromatic cage residues, and therefore may disrupt its binding to H3K36me2/3 (Figure 2B).

In TBRS patients, mutations in the ADD domain cluster around the H3K4 binding site. Crystal structure of DNMT3A bound to H3K4me0 shows that Asp529 in the ADD domain makes direct contact with the Lys4 of the histone protein H3 [56]. In absence of the histone H3 peptide, Asp529 also interacts with the DNA binding region, suggesting its involvement in regulation of the autoinhibited state. The TBRS variant Asp529Asn therefore could potentially be hyperactive and mistargeted. Biochemical data showing that the variant Asp529Ala is neither autoinhibited by the ADD domain nor activated by unmethylated H3 peptide support this speculation [56]. Met548 and Trp581 also make direct contact with the H3 tail, and therefore the variants Met548Thr and Trp581Cys may have reduced interaction with histone H3. Indeed, as reported, Met548Trp variant cannot be released from the autoinhibited state in the presence of unmethylated H3 peptide [56]. Besides these mutations Cys549, Cys562, and Cys583 are present in the zinc finger regions. These residues mediate zinc binding and the mutations could potentially alter the DNMT3A structure or stability.

In TBRS patients, 15 substitutions and 2 frameshift mutations in the MTase domain are distributed in all sequence motifs (Figure 3A) [65,66]. We analyzed the crystal structure of DNMT3A to predict the potential effects of TBRS mutations on its activity (Figure 3B). As visualized in the structure of DNMT3A, Trp698, Pro700 and Arg736, are spatially located near Motifs I–III, and therefore the mutations at these residues could affect the AdoMet binding and catalysis (Figure 3C). Pro700 interacts with Arg635 in Motif I, and Arg736 contacts the backbones of Arg688 in Motif III (Figure 3C). The most recent co-crystal structure of DNMT3A bound to DNA shows that Ser714 interacts with phosphodiester backbone and Arg749 interacts with Asp702 of motif IV, which is involved in catalysis (Figure 3D) [74]. Biochemical investigation of the TBRS substitutions Arg736His and Ser714Cys were recently reported [22]. The data show that Arg736His variant, interestingly, has a 3-fold increase in catalytic turnover but is weakly stimulated by DNMT3L, and has an increased preference for non-CpG sites. However, Ser714Cys has reduced activity as well as reduced stimulation by DNMT3L [22,74]. These data suggest that besides having a direct effect on catalysis, these mutations could alter the structure disrupting the interaction of DNMT3A with DNMT3L. The regions where DNMT3A interacts with the DNA (TRD) spans from motif VIII–IX. Residues that are mutated in or near this region include Val778, Met801, Asn838, Arg882, Phe902, and Pro904. Val778 and Met801 are spatially located near motif VIII, so their mutations to Gly and Val, respectively, may alter DNMT3A's ability to bind to DNA (Figure 3E). Asn838 interacts with the phosphodiester bond between nucleotides at N+2 and N+3 from the target CpG site, suggesting that the Asn838Asp variant may have a weak binding to DNA (Figure 3F) [59,74]. Phe902, Pro904 and Leu648 are spatially located near motif X (Figure 3G). In vitro studies show that Pro904Leu variant has higher catalytic turnover and negligible effect on DNMT3L mediated DNMT3A stimulation [22].

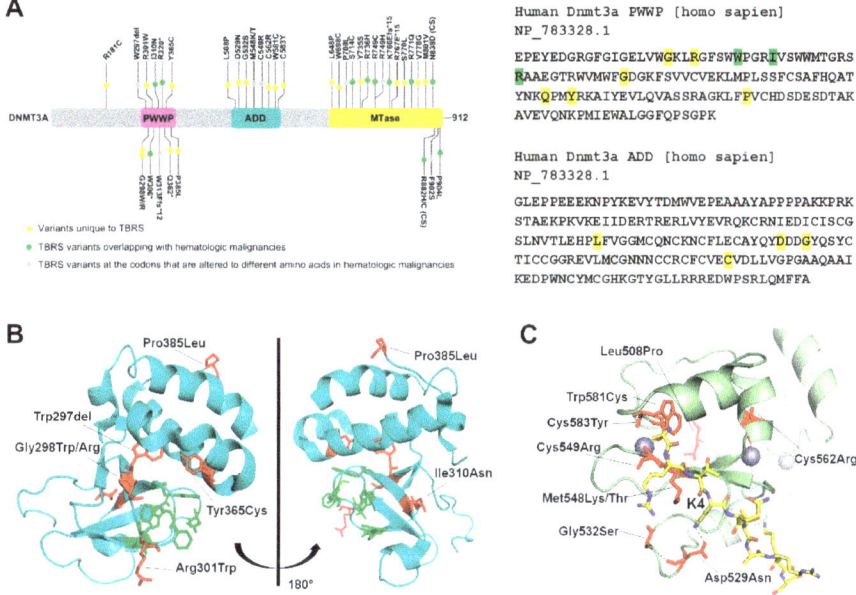

Figure 2. TBRS (Tatton-Brown-Rahman syndrome) mutations in the PWWP and ADD domain of DNMT3A (**A**) Left Schematic representation of the hDNMT3A gene, with mutations listed in different domains. Variants unique to TBRS are highlighted in yellow, TBRS variants overlapping with hematologic malignancies are highlighted in green, and TBRS variants at the codon that are altered to different amino acids in hematologic malignancies are highlighted in grey. In the schematic, (*) is used to indicate a stop codon replacement, Ffs indicates a frame-shift mutation, and CS indicates the catalytic site. Right Nucleotide sequence of the PWWP (Pro-Trp-Trp-Pro) and ADD (ATRX-DNMT3-DNMT3L) domain, with the mutations highlighted in the color corresponding to the schematic. (**B**) Two orientations of the DNMT3A PWWP domain (PDB: 3LLR). The positions of TBRS mutations are shown as stick structures in red, whereas residues part of the aromatic cage that bind to H3K36me2/3 shown as stick structures in green. (**C**) Crystal Structure of the DNMT3A ADD domain (PDB: 4U7T), in green bound to an unmodified H3 peptide shown as a stick structure in yellow. Grey spheres represent bound zinc. The positions of TBRS mutations are shown as a stick structure in red.

The TBRS variant, Arg882His, which is also the most prominent somatic variant of DNMT3A in acute myeloid leukemia (AML) patients, has been extensively studied (Figure 3F) [20]. Interestingly mutation of Arg882 is found in 25% AML patients and 25% of TBRS patients indicating this to be a hotspot [72]. Given the majority of TBRS patients are pediatric or young adults, it is difficult to determine the risk of AML, because of its late onset. More recently, two TBRS patients were diagnosed with AML in childhood supporting potential risk and susceptibility of TBRS patients to develop AML [66].

The effect of Arg882His substitution on DNMT3A activity, reported by several studies, caused a 40–80% loss of catalytic activity [75,76]. Arg882 interacts with the phosphodiester bond of the nucleotide at N+3 from the target CpG site. Given the position of R882 close to the RD interface, the mutation disrupts intermolecular interactions, thus preventing the enzyme to form tetramers (Figure 3F). This in turn negatively affects DNMT3A cooperativity and decreases its DNA binding capacity [23]. More recently, the variant was shown to have altered flanking sequence preference around the CpG site [25]. However, it is not clear whether altered flanking sequence preference is a direct consequence of the amino acid substitution or an indirect effect of the loss of cooperativity. Based on its effect on genomic DNA methylation, Arg882His is also suggested to have a dominant-negative effect on the wild type enzyme, however this activity was not confirmed by in vitro experiments [25,76,77].

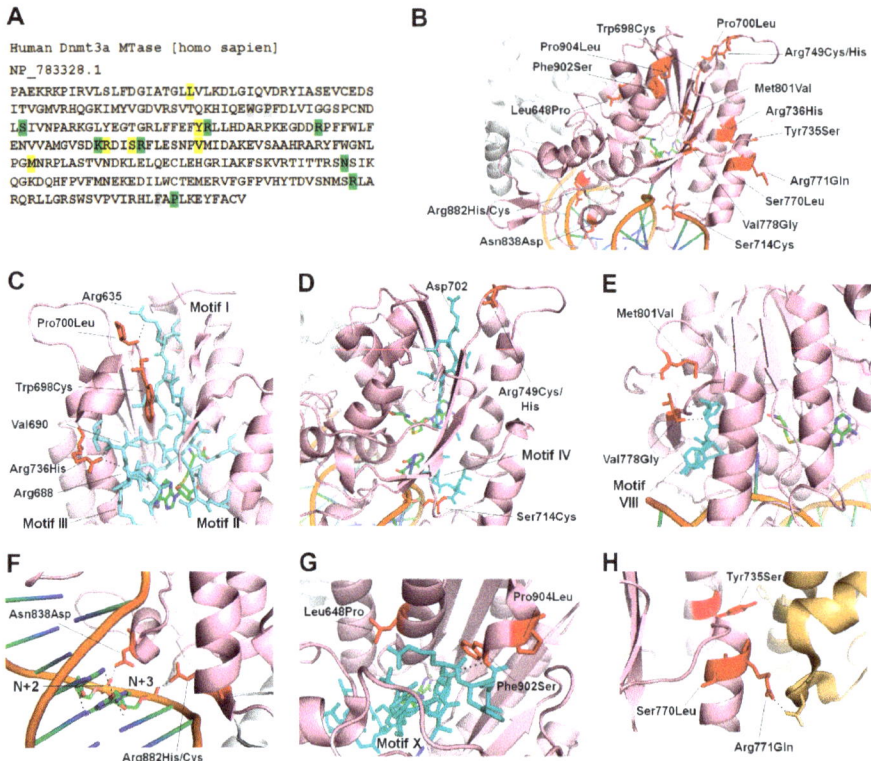

Figure 3. TBRS mutations in the catalytic domain of DNMT3A (A) Nucleotide sequence of the hDNMT3A MTase domain. As seen in Figure 2A, variants unique to TBRS are highlighted in yellow, TBRS variants overlapping with hematologic malignancies are highlighted in green, and TBRS variants at the codon that are altered to different amino acids in hematologic malignancies are highlighted in grey. (B) Crystal structure of DNMT3A bound to DNA, zoomed in to show only one of the two monomers in the tetrameric structure (PDB: 5YX2). The MTase domain is shown in a pink cartoon structure and the positions of TBRS mutations are shown as stick structures in red. The second monomer of the MTase is shown in grey. (C–G) Magnified view of the motifs I-III (C), motif IV (D–E), motif VIII (F), the TRD (G), and motif X (H) shown as stick structures in blue and the positions of TBRS mutations shown as stick structures in red. The black dotted lines represent interactions with nearby residues or with the DNA. (H) The interface between DNMT3A, pink cartoon, and DNMT3L, orange cartoon. The positions of TBRS mutations shown as stick structures in red. The black dotted lines represent interactions with nearby residues of DNMT3L shown as stick structures in orange.

Another group of TBRS mutations includes residues Tyr735, Ser770, and Arg771, involved in intermolecular interactions with DNMT3L (Figure 3H). Tyr735 and Arg771 make direct contact with His and Asp residues in DNMT3L, respectively, so their mutation may alter the stability of this interface.

Biochemical analysis of the Arg771Gln variant interestingly shows an increase in the catalytic activity of the enzyme by 6-fold and no change in the level of stimulation by DNMT3L [22]. However other substitutions of this residue resulted in a decrease in the stimulation by DNMT3L showing the role of this residue in stabilization of DNMT3A-3L interactions.

It is interesting to note that while majority of the mutated residues are highly conserved in DNMT3B, three residues Leu648, Ser714, and Arg736 are conserved in bacterial DNA cytosine methyltransferases including M. HhaI and M. HaeIII, suggesting the effect of substitutions on their conserved structure and catalytic mechanism. However, compared to the level of overlap between DNMT3A mutations found in hematologic malignancies and growth syndromes, very few coincide with mutations of DNMT3B found in ICF patients. Further, the effect of TBRS mutation, Arg736His, on the activity of DNMT3A is notable given that at this position His is normally present in DNMT3B. This suggests that Arg736 in DNMT3A is important for the catalytic mechanism, which has common and distinct features from that of DNMT3B. These observations suggest that the catalytic mechanisms of DNMT3A and DNMT3B are critical for their unique biological functions. While this speculation is anticipated, effects of these substitutions on the catalytic mechanism of DNMT3A compared to DNMT3B need to be further elucidated.

3.2. Hereditary Tumors and Microcephalic Dwarfism (MD)

Pheochromocytoma/paraganglioma (PCC/PGL) is a rare neuroendocrine malignancy that may develop at various body sites, including the head, neck, and abdomen, and has a five-year survival rate of only 40% [78,79]. PCC/PGL is the most heritable of all tumors and carries both germline and somatic mutations in 1 of 20 known genes including metabolic genes. Recently de novo germline mutations in DNMT3A were reported in PCC/PGLs. The mutations occur in the PWWP domain, and result in substitutions Lys299Ile and Arg318Trp (Figure 4A). Although these residues are not the part of the aromatic cage, crystal structure analysis shows that Lys299 interacts with the backbone of Phe303, which stabilizes the interaction of PWWP domain with H3K36me2/3 (Figure 4B). Conversely, Arg318 interacts with the Val35 of the H3 tail, so the substitution Arg318Trp may effect H3 binding. However, future biochemical studies will be needed to show the effect of these mutations on the activity of the enzymes. Methylation profiling of PCC/PGL patients and of HeLa cells carrying CRISPR/Cas9-mediated knock-in of the PGG/PCL DNMT3A mutation show site-specific hypermethylation at homeobox genes, genes involved in dopaminergic neurogenesis, neural crest differentiation, and embryonic morphogenesis. Given that previously known mutations in DNMT3A causing overgrowth syndrome result in genome-wide hypomethylation, the PWWP mutation in PCC/PGL leading to hypermethylation is described as gain-of-function mutation.

Similar DNMT3A heterozygous gain-of-function mutations were recently shown to cause microcephalic dwarfism, a hypocellular disorder of extreme global growth failure, including a reduction in head size and height [80]. The mutations result in substitutions Trp330Arg and Asp333Asn, both of which are located in the DNMT3A PWWP domain (Figure 4C). Both residues are part of the aromatic cage that interacts with H3K36me2/3, and when mutated, the interaction with chromatin is abrogated (Figure 4D) [80]. In contrast to some TBRS mutations, which lead to PWWP domain instability, MD mutations do not affect the stability of this domain [80]. Genome-wide DNA methylation analysis of patients' fibroblasts showed that similar to PCC/PGL, the majority of differentially methylated regions (DMRs) were hypermethylated compared to wild type samples, and these DMRs were associated with developmental transcription factors and morphogen genes. In the control fibroblasts, these regions were marked by tri-methylation of lysine 27 on histone H3 (H3K27), which is established by the polycomb repressive complex, PRC2, and DNA remains unmethylated [80]. Notably, many of the hypermethylated regions in patient fibroblasts were identified as broad non-methylated islands or differentially methylated valleys in normal cells [81–84]. Reduced H3K27me3 at these hypermethylated sites occurs despite normal levels of the PRC2 subunit, EZH2 histone methyltransferases in the MD patients. Given that MD DNMT3A variants cannot be targeted to H3K36me3 chromatin domains, they are speculated to methylate transcriptionally repressed regions spuriously, which are otherwise regulated by the PRC2 complex. This phenomenon could be explained by previous observations showing an antagonism between DNA methylation and deposition of H3K27me3. It was further shown that DNA methylation can abrogate binding of the PRC2 protein, SUZ12, to nucleosomes that can impact EZH2 activity at these sites [82,85]. These observations support the speculation that PRC2 binding is impaired by DNA methylation at hypermethylated regions in MD patients. Whereas, loss of EZH2 promotes premature differentiation, loss of DNMT3A increases the stemness of embryonic and progenitor stem cells [86]. The DNMT3A MD variants, however, are proposed to be gain-of-function mutations, which might increase cellular differentiation causing premature depletion of stem/progenitor cell pools. This in turn could affect the growth of tissues and lead to reduced organism size [80].

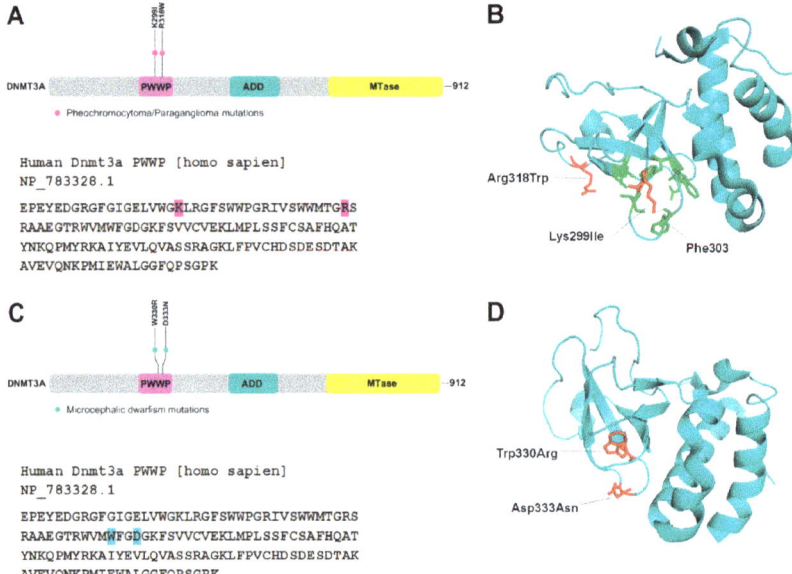

Figure 4. PCC/PGL (Pheochromocytoma/paraganglioma) and MD (microcephalic dwarfism) mutations in PWWP domain of DNMT3A (**A**,**C**) Top Schematic representation of the *hDNMT3A* gene, showing PCC/PGL and MD mutations in the PWWP domain. Below Nucleotide sequence of the PWWP domain, with the mutations highlighted in pink and blue respectively. (**B**,**D**) Crystal structure of DNMT3A PWWP domain shown in blue (PDB: 3LLR). The positions of PCC/PGL mutation (**B**) and MD mutations (**D**) shown as stick structures in red. The aromatic cage residues are shown in green.

Comparatively, a single mutation in the PWWP domain of DNMT3B was discovered in patients with ICF syndrome [87]. The mutation S282P also resides in the hydrophobic pocket that interacts with H3K36me3. Earlier data showed that that this mutation leads to decreased or loss of heterochromatin binding of DNMT3B [88]. Later was also shown to affect the interaction of DNMT3B with H3K36me3 at gene bodies [53,89]. Given the complexity imposed by various alternatively spliced isoforms of DNMT3B on its biological activity in various tissues, the S282P variant could have a more diverse effect on DNA methylation landscape compared to similar mutation in DNMT3A. Indeed, the catalytically inactive isoforms of DNMT3B were shown mediate DNA methylation at gene bodies by potentially recruiting DNMT3A to these sites, therefore demonstrating a functional role similar to DNMT3L [90,91]. Global expression and epigenetic profiling of cells derived from ICF patients showed upregulation of genes associated accompanied by loss of SUZ12 binding and H3K27me3. However, in contrast to the growth syndrome MD, these regions remain hypomethylated [92]. This distinctive feature could be due to specific recruitment, protein-protein interactions and activities of DNMT3A and DNMT3B.

4. Conclusions and Perspectives

DNA methylation defects in the absence of DNMT mutations have been reported in a plethora of disorders. However, with exception of DNMT3B germline transmitted mutations that cause ICF, disease-causing mutations in DNMT1 and DNMT3A were recently discovered. These mutations cause growth disorders, HSAN1E, ADCA-DN, TBRS and MD that have distinct clinical manifestations.

The ADCA-DN and HSAN1E mutations were mapped to the regions that could potentially impair the interaction of DNMT1-RFTS domain either with the ubiquitin or with the DNA binding region of the MTase domain. The former will render the enzyme in an autoinhibited state, and the latter will lead to an unregulated activity of DNMT1 enzyme making it hyperactive (Figure 1D). Methylome mapping of the patients show aberrant DNA methylation including both widespread hypomethylation and site-specific hypermethylation. This methylation pattern is similar to the one observed in various cancers indicating a similar loss of DNMT1 function, however none of the HSAN1E and ADCA-DN patients develop cancer. It is speculated that in these patients, a gradual loss of DNA methylation over time may cause late onset and progressive neurological disabilities. However, given that post mitotic neurons are terminally differentiated and do not perform maintenance methylation, the effect of RFTS mutations on the activity of DNMT1 could be due to protein misfolding and aggregation [30]. Interestingly, ADCA-DN and HSN1E have some overlapping clinical features typical of mitochondrial diseases, which is supported by biochemical evidence of mitochondrial dysfunction [33,37,93]. Given that presence of methyl cytosine in mtDNA is still debated, it is not clear how DNMT1 mutations can cause mitochondrial dysfunction through its effect on mtDNA methylation. Taken together, these data suggest that DNMT1 mutants could exert their damaging effect through at least two mechanisms including impairment of epigenetic pathways, and cellular stress by the protein misfolding [30,32,33].

Compared to TBRS in which DNMT3A mutations are present in all functional domains of the protein, in MD, the two mutations map only to the PWWP domain. However, in contrast to MD-specific PWWP mutations, the TBRS-specific PWWP mutations affected protein stability resulting in loss of protein function. Based on the observation that the MD variant, DNMT3A Trp330Arg, is unable to interact with H3K36me2/3, it was proposed to be "available" to methylate sites that are normally polycomb repressed [80]. This is consistent with the observation that polycomb repressed developmental genes in ESCs gain DNA methylation during differentiation and are often found hypermethylated in cancer cells [12,83,94]. It is possible that by interacting with PRC2, the DNMT3A Trp330Arg variant is targeted to these regions in MD patients [81,84]. In addition, once the DNA methylation is established, it could potentially block the activity of PRC2 by interfering with the binding of Suz12 [82] and result in stable repression of the developmental genes [85]. Comparatively, in ICF cells, the DNMT3B Ser282Pro variant, which also has impaired binding to H3K36me3, does not methylate PRC2 repressed regions. However, it is obscure how these regions lose PRC2 binding and H3k27me3. Furthermore, mutations of histone methyltransferases, EZH2 and NSD1, cause the Weaver and Sotos overgrowth syndromes, respectively, which supports a critical role of epigenetic regulation in organism size.

The discovery of DNMT mutations in rare diseases has shaped our understanding of the cause and consequence of aberrant DNA methylation in various disorders. It is clear that global hypomethylation, often found in cancers as well as growth disorders, may not be a direct consequence of DNMT loss of function, rather an indirect response to a diseased state. Targeted hypermethylation seems to be direct consequence of aberrant DNMT activity. This is also supported by inconsistency between the biochemical outcomes of DNMT mutations and their effect on genomic methylation. Future studies designed to address the direct effect of DNMT mutations on genomic DNA methylation patterns will help understand the contribution of DNMTs in pathogenesis.

Author Contributions: All authors contributed to the writing and correction of the review.

Funding: Work in Gowher lab is supported by NIHR01GM118654-01 and NSF 1,716,678 grants.

Acknowledgments: We are thankful to Gowher lab members for discussions.

Conflicts of Interest: The authors declare no conflict of interest.

References

1. Goll, M.G.; Bestor, T.H. Eukaryotic cytosine methyltransferases. *Annu. Biochem.* **2005**, *74*, 481–514. [CrossRef]
2. Edwards, J.R.; Yarychkivska, O.; Boulard, M.; Bestor, T.H.; Boulard, M. DNA methylation and DNA methyltransferases. *Epigenetics Chromatin* **2017**, *10*, 23. [CrossRef]
3. Jones, P.A. Functions of DNA methylation: Islands, start sites, gene bodies and beyond. *Nat. Rev. Microbiol.* **2012**, *13*, 484–492. [CrossRef]
4. Wu, X.; Zhang, Y. TET-mediated active DNA demethylation: Mechanism, function and beyond. *Nat. Rev. Microbiol.* **2017**, *18*, 517–534. [CrossRef]
5. Wu, H.; Zhang, Y. Reversing DNA methylation: Mechanisms, genomics, and biological functions. *Cell* **2014**, *156*, 45–68. [CrossRef] [PubMed]
6. Li, J.; Wu, X.; Zhou, Y.; Lee, M.; Guo, L.; Han, W.; Mo, W.; Cao, W.-M.; Sun, D.; Xie, R.; et al. Decoding the dynamic DNA methylation and hydroxymethylation landscapes in endodermal lineage intermediates during pancreatic differentiation of hESC. *Nucleic Acids* **2018**, *46*, 2883–2900. [CrossRef] [PubMed]
7. Mahé, E.A.; Madigou, T.; Sérandour, A.A.; Bizot, M.; Avner, S.; Chalmel, F.; Palierne, G.; Métivier, R.; Salbert, G. Cytosine modifications modulate the chromatin architecture of transcriptional enhancers. *Genome Res.* **2017**, *27*, 947–958. [CrossRef]
8. Hackett, J.A.; Dietmann, S.; Murakami, K.; Down, T.A.; Leitch, H.G.; Surani, M.A. Synergistic mechanisms of DNA demethylation during transition to ground-state pluripotency. *Stem Cell Rep.* **2013**, *1*, 518–531. [CrossRef]
9. Cheng, X. Structure and function of DNA methyltransferases. *Annu. Biophys. Biomol. Struct.* **1995**, *24*, 293–318. [CrossRef]
10. Gowher, H.; Jeltsch, A. Mammalian DNA methyltransferases: New discoveries and open questions. *Biochem. Soc. Trans.* **2018**, *46*, 1191–1202. [CrossRef] [PubMed]
11. Bestor, T.H. The DNA methyltransferases of mammals. *Hum. Mol. Genet.* **2000**, *9*, 2395–2402. [CrossRef] [PubMed]
12. Mortusewicz, O.; Schermelleh, L.; Walter, J.; Cardoso, M.C.; Leonhardt, H. Recruitment of DNA methyltransferase I to DNA repair sites. *Proc. Natl. Acad. Sci. USA* **2005**, *102*, 8905–8909. [CrossRef] [PubMed]
13. Chen, T.; Li, E. Establishment and maintenance of DNA methylation patterns in mammals. In *DNA Methylation: Basic Mechanisms*; Springer: Berlin/Heidelberg, Germany, 2006; pp. 179–201.
14. Gowher, H.; Liebert, K.; Hermann, A.; Xu, G.; Jeltsch, A. Mechanism of stimulation of catalytic activity of Dnmt3A and Dnmt3B DNA-(cytosine-C5)-methyltransferases by Dnmt3L. *J. Boil. Chem.* **2005**, *280*, 13341–13348. [CrossRef] [PubMed]
15. Okano, M.; Bell, D.W.; Haber, D.A.; Li, E. DNA methyltransferases Dnmt3a and Dnmt3b are essential for de novo methylation and mammalian development. *Cell* **1999**, *99*, 247–257. [CrossRef]
16. Lyko, F. The DNA methyltransferase family: A versatile toolkit for epigenetic regulation. *Nat. Rev. Genet.* **2018**, *19*, 81–92. [CrossRef] [PubMed]
17. Jurkowska, R.Z.; Jeltsch, A. Enzymology of mammalian DNA methyltransferases. *Purine Pyrimidine Metab. Man V* **2016**, *945*, 87–122.
18. Tajima, S.; Suetake, I.; Takeshita, K.; Nakagawa, A.; Kimura, H. Domain structure of the Dnmt1, Dnmt3a, and Dnmt3b DNA methyltransferases. *Adv. Exp. Med. Boil.* **2016**, *945*, 63–86.
19. Jones, P.A.; Issa, J.-P.J.; Baylin, S. Targeting the cancer epigenome for therapy. *Nat. Rev. Microbiol.* **2016**, *17*, 630–641. [CrossRef]
20. Brunetti, L.; Gundry, M.C.; Goodell, M.A. DNMT3A in Leukemia. *Cold Spring Harb. Perspect. Med.* **2017**, *7*. [CrossRef]

21. Velasco, G.; Francastel, C. Genetics meets DNA methylation in rare diseases. *Clin. Genet.* **2019**, *95*, 210–220. [CrossRef]
22. Sandoval, J.E.; Huang, Y.H.; Muise, A.; Goodell, M.A.; Reich, N.O. Mutations in the DNMT3A DNA methyltransferase in AML patients cause both loss and gain of function and differential regulation by protein partners. *J. Biol. Chem.* **2019**. [CrossRef] [PubMed]
23. Norvil, A.B.; Petell, C.J.; Alabdi, L.; Wu, L.; Rossie, S.; Gowher, H. Dnmt3b methylates DNA by a noncooperative mechanism, and its activity is unaffected by manipulations at the predicted dimer interface. *Biochemistry* **2016**, *57*, 4312–4324. [CrossRef]
24. Emperle, M.; Dukatz, M.; Kunert, S.; Holzer, K.; Rajavelu, A.; Jurkowska, R.Z.; Jeltsch, A. The DNMT3A R882H mutation does not cause dominant negative effects in purified mixed DNMT3A/R882H complexes. *Sci. Rep.* **2018**, *8*, 13242. [CrossRef] [PubMed]
25. Emperle, M.; Rajavelu, A.; Kunert, S.; Arimondo, P.B.; Reinhardt, R.; Jurkowska, R.Z.; Jeltsch, A. The DNMT3A R882H mutant displays altered flanking sequence preferences. *Nucleic Acids* **2018**, *46*, 3130–3139. [CrossRef] [PubMed]
26. Holz-Schietinger, C.; Matje, D.M.; Reich, N.O. Mutations in DNA methyltransferase (DNMT3A) observed in acute myeloid leukemia patients disrupt processive methylation. *J. Boil. Chem.* **2012**, *287*, 30941–30951. [CrossRef] [PubMed]
27. Song, J.; Rechkoblit, O.; Bestor, T.H.; Patel, D.J. Structure of DNMT1-DNA complex reveals a role for autoinhibition in maintenance DNA methylation. *Science* **2011**, *331*, 1036–1040. [CrossRef]
28. Bostick, M.; Kim, J.K.; Estève, P.-O.; Clark, A.; Pradhan, S.; Jacobsen, S.E. UHRF1 plays a role in maintaining DNA methylation in mammalian cells. *Science* **2007**, *317*, 1760–1764. [CrossRef]
29. Svedruzic, Z.M. Dnmt1 structure and function. *Prog. Mol. Biol. Transl. Sci.* **2011**, *101*, 221–254. [PubMed]
30. Baets, J.; Duan, X.; Wu, Y.; Smith, G.; Mademan, I.; Khoury, J.; Botuyan, M.-V.; Mer, G.; Hojo, K.; DeLeon, J.; et al. Defects of mutant DNMT1 are linked to a spectrum of neurological disorders. *Brain* **2015**, *138*, 845–861. [CrossRef]
31. Winkelmann, J.; Lin, L.; Schormair, B.; Kornum, B.R.; Faraco, J.; Plazzi, G.; Melberg, A.; Cornelio, F.; Urban, A.E.; Pizza, F.; et al. Mutations in DNMT1 cause autosomal dominant cerebellar ataxia, deafness and narcolepsy. *Hum. Mol. Genet.* **2012**, *21*, 2205–2210. [CrossRef]
32. Klein, C.J.; Botuyan, M.-V.; Wu, Y.; Ward, C.J.; Nicholson, G.A.; Hammans, S.; Hojo, K.; Yamanishi, H.; Karpf, A.R.; Wallace, D.C.; et al. Mutations in DNMT1 cause hereditary sensory neuropathy with dementia and hearing loss. *Nat. Genet.* **2011**, *43*, 595–600. [CrossRef]
33. Sun, Z.; Wu, Y.; Ordog, T.; Baheti, S.; Nie, J.; Duan, X.; Hojo, K.; Kocher, J.-P.; Dyck, P.J.; Klein, C.J. Aberrant signature methylome by DNMT1 hot spot mutation in hereditary sensory and autonomic neuropathy 1E. *Epigenetics* **2014**, *9*, 1184–1193. [CrossRef]
34. Wu, B.-K.; Mei, S.-C.; Brenner, C. RFTS-deleted DNMT1 enhances tumorigenicity with focal hypermethylation and global hypomethylation. *Cell Cycle* **2014**, *13*, 3222–3231. [CrossRef]
35. Smets, M.; Link, S.; Wolf, P.; Schneider, K.; Solis, V.; Ryan, J.; Meilinger, D.; Qin, W.; Leonhardt, H. DNMT1 mutations found in HSANIE patients affect interaction with UHRF1 and neuronal differentiation. *Hum. Mol. Genet.* **2017**, *26*, 1522–1534. [CrossRef]
36. Klein, C.J.; Bird, T.; Ertekin-Taner, N.; Lincoln, S.; Hjorth, R.; Wu, Y.; Kwok, J.; Mer, G.; Dyck, P.J.; Nicholson, G.A. *DNMT1* mutation hot spot causes varied phenotypes of HSAN1 with dementia and hearing loss. *Neurology* **2013**, *80*, 824–828. [CrossRef]
37. Moghadam, K.K.; Pizza, F.; La Morgia, C.; Franceschini, C.; Tonon, C.; Lodi, R.; Barboni, P.; Seri, M.; Ferrari, S.; Liguori, R.; et al. Narcolepsy is a common phenotype in HSAN IE and ADCA-DN. *Brain* **2014**, *137*, 1643–1655. [CrossRef]
38. Yuan, J.; Higuchi, Y.; Nagado, T.; Nozuma, S.; Nakamura, T.; Matsuura, E.; Hashiguchi, A.; Sakiyama, Y.; Yoshimura, A.; Takashima, H. Novel mutation in the replication focus targeting sequence domain of DNMT1 causes hereditary sensory and autonomic neuropathy IE. *J. Peripher. Nerv.* **2013**, *18*, 89–93. [CrossRef]
39. Zhang, Z.-M.; Liu, S.; Lin, K.; Luo, Y.; Perry, J.J.; Wang, Y.; Song, J. Crystal structure of human DNA methyltransferase 1. *J. Mol. Boil.* **2015**, *427*, 2520–2531. [CrossRef]
40. Li, T.; Wang, L.; Du, Y.; Xie, S.; Yang, X.; Lian, F.; Zhou, Z.; Qian, C. Structural and mechanistic insights into UHRF1-mediated DNMT1 activation in the maintenance DNA methylation. *Nucleic Acids* **2018**, *46*, 3218–3231. [CrossRef]

41. Pedroso, J.L.; Barsottini, O.G.P.; Lin, L.; Melberg, A.; Oliveira, A.S.B.; Mignot, E. A Novel de novo exon 21 DNMT1 mutation causes cerebellar ataxia, deafness, and narcolepsy in a Brazilian patient. *Sleep* **2013**, *36*, 1257–1259. [CrossRef]
42. Kernohan, K.D.; Care4Rare Canada Consortium; Schenkel, L.C.; Huang, L.; Smith, A.; Pare, G.; Ainsworth, P.; Boycott, K.M.; Warman-Chardon, J.; Sadikovic, B. Identification of a methylation profile for DNMT1-associated autosomal dominant cerebellar ataxia, deafness, and narcolepsy. *Clin. Epigenetics* **2016**, *8*, 91. [CrossRef]
43. Mastroeni, D.; Grover, A.; Delvaux, E.; Whiteside, C.; Coleman, P.D.; Rogers, J. Epigenetic changes in Alzheimer's disease: Decrements in DNA methylation. *Neurobiol. Aging* **2010**, *31*, 2025–2037. [CrossRef]
44. Yokoyama, A.S.; Rutledge, J.C.; Medici, V. DNA methylation alterations in Alzheimer's disease. *Environ. Epigenetics* **2017**, *3*, 008. [CrossRef]
45. Jowaed, A.; Schmitt, I.; Kaut, O.; Wüllner, U. Methylation regulates α-synuclein expression and is decreased in Parkinson's disease patients' brains. *J. Neurosci.* **2010**, *30*, 6355–6359. [CrossRef]
46. Desplats, P.; Spencer, B.; Coffee, E.; Patel, P.; Michael, S.; Patrick, C.; Adame, A.; Rockenstein, E.; Masliah, E. α-Synuclein sequesters Dnmt1 from the nucleus a novel mechanism for epigenetic alterations in lewy body diseases. *J. Biol. Chem.* **2011**, *286*, 9031–9037. [CrossRef]
47. Wüllner, U.; Kaut, O.; deBoni, L.; Piston, D.; Schmitt, I. DNA methylation in Parkinson's disease. *J. Neurochem.* **2016**, *139*, 108–120. [CrossRef]
48. Weissman, J.; Naidu, S.; Bjornsson, H.T. Abnormalities of the DNA methylation mark and its machinery: An emerging cause of neurologic dysfunction. *Semin. Neurol.* **2014**, *34*, 249–257. [CrossRef]
49. Kanai, Y.; Ushijima, S.; Nakanishi, Y.; Sakamoto, M.; Hirohashi, S. Mutation of the *DNA methyltransferase (DNMT)* 1 gene in human colorectal cancers. *Cancer Lett.* **2003**, *192*, 75–82. [CrossRef]
50. Forbes, S.; Bhamra, G.; Bamford, S.; Dawson, E.; Kok, C.; Clements, J.; Menzies, A.; Teague, J.; Futreal, P.; Stratton, M. The Catalogue of Somatic Mutations in Cancer (COSMIC). *Curr. Protoc. Hum. Genet.* **2008**, *57*, Unit 10.11.
51. Cheng, X.; Blumenthal, R.M. Mammalian DNA methyltransferases: A structural perspective. *Structure* **2008**, *16*, 341–350. [CrossRef]
52. Chen, T.; Tsujimoto, N.; Li, E. The PWWP domain of Dnmt3a and Dnmt3b is required for directing DNA methylation to the major satellite repeats at pericentric heterochromatin. *Mol. Cell. Boil.* **2004**, *24*, 9048–9058. [CrossRef] [PubMed]
53. Dhayalan, A.; Rajavelu, A.; Rathert, P.; Tamas, R.; Jurkowska, R.Z.; Ragozin, S.; Jeltsch, A. The Dnmt3a PWWP domain reads histone 3 Lysine 36 trimethylation and guides DNA methylation. *J. Boil. Chem.* **2010**, *285*, 26114–26120. [CrossRef]
54. Wu, H.; Zeng, H.; Lam, R.; Tempel, W.; Amaya, M.F.; Xu, C.; Dombrovski, L.; Qiu, W.; Wang, Y.; Min, J. Structural and histone binding ability characterizations of human PWWP domains. *PLoS ONE* **2011**, *6*, e18919. [CrossRef] [PubMed]
55. Ooi, S.K.T.; Qiu, C.; Bernstein, E.; Li, K.; Jia, D.; Yang, Z.; Erdjument-Bromage, H.; Tempst, P.; Lin, S.-P.; Allis, C.D.; et al. DNMT3L connects unmethylated lysine 4 of histone H3 to de novo methylation of DNA. *Nat. Cell Boil.* **2007**, *448*, 714–717. [CrossRef]
56. Guo, X.; Wang, L.; Li, J.; Ding, Z.; Xiao, J.; Yin, X.; He, S.; Shi, P.; Dong, L.; Li, G.; et al. Structural insight into autoinhibition and histone H3-induced activation of DNMT3A. *Nature* **2015**, *517*, 640. [CrossRef] [PubMed]
57. Petell, C.J.; Alabdi, L.; He, M.; Miguel, P.S.; Rose, R.; Gowher, H. An epigenetic switch regulates de novo DNA methylation at a subset of pluripotency gene enhancers during embryonic stem cell differentiation. *Nucleic Acids* **2016**, *44*, 7605–7617. [CrossRef] [PubMed]
58. Gowher, H.; Loutchanwoot, P.; Vorobjeva, O.; Handa, V.; Jurkowska, R.Z.; Jurkowski, T.P.; Jeltsch, A. Mutational analysis of the catalytic domain of the murine Dnmt3a DNA-(cytosine C5)-methyltransferase. *J. Mol. Boil.* **2006**, *357*, 928–941. [CrossRef]
59. Rajavelu, A.; Jurkowska, R.Z.; Fritz, J.; Jeltsch, A. Function and disruption of DNA Methyltransferase 3a cooperative DNA binding and nucleoprotein filament formation. *Nucleic Acids* **2011**, *40*, 569–580. [CrossRef]
60. Jia, D.; Jurkowska, R.Z.; Zhang, X.; Jeltsch, A.; Cheng, X. Structure of Dnmt3a bound to Dnmt3L suggests a model for de novo DNA methylation. *Nat. Cell Boil.* **2007**, *449*, 248–251. [CrossRef]
61. Emperle, M.; Rajavelu, A.; Reinhardt, R.; Jurkowska, R.Z.; Jeltsch, A. Cooperative DNA binding and protein/DNA fiber formation increases the activity of the Dnmt3a DNA methyltransferase. *J. Boil. Chem.* **2014**, *289*, 29602–29613. [CrossRef]

62. Hansen, R.S.; Wijmenga, C.; Luo, P.; Stanek, A.M.; Canfield, T.K.; Weemaes, C.M.R.; Gartler, S.M. The *DNMT3B* DNA methyltransferase gene is mutated in the ICF immunodeficiency syndrome. *Proc. Acad. Sci.* **1999**, *96*, 14412–14417. [CrossRef]
63. Xu, G.-L.; Bestor, T.H.; Bourc'His, D.; Hsieh, C.-L.; Tommerup, N.; Bugge, M.; Hulten, M.; Qu, X.; Russo, J.J.; Viegas-Péquignot, E. Chromosome instability and immunodeficiency syndrome caused by mutations in a DNA methyltransferase gene. *Nat. Cell Boil.* **1999**, *402*, 187–191. [CrossRef]
64. Ehrlich, M. The ICF syndrome, a DNA methyltransferase 3B deficiency and immunodeficiency disease. *Clin. Immunol.* **2003**, *109*, 17–28. [CrossRef]
65. Tatton-Brown, K.; Childhood Overgrowth Consortium; Seal, S.; Ruark, E.; Harmer, J.; Ramsay, E.; Duarte, S.D.V.; Zachariou, A.; Hanks, S.; O'Brien, E.; et al. Mutations in the DNA methyltransferase gene *DNMT3A* cause an overgrowth syndrome with intellectual disability. *Nat. Genet.* **2014**, *46*, 385–388. [CrossRef] [PubMed]
66. Tatton-Brown, K.; Zachariou, A.; Loveday, C.; Renwick, A.; Mahamdallie, S.; Aksglaede, L.; Baralle, D.; Barge-Schaapveld, D.; Blyth, M.; Bouma, M.; et al. The Tatton-Brown-Rahman syndrome: A clinical study of 55 individuals with de novo constitutive DNMT3A variants. *Wellcome Open* **2018**, *3*, 46. [CrossRef]
67. Okamoto, N.; Toribe, Y.; Shimojima, K.; Yamamoto, T. Tatton-Brown-Rahman syndrome due to 2p23 microdeletion. *Am. J. Med. Genet. Genet. A* **2016**, *170*, 1339–1342. [CrossRef]
68. Hollink, I.H.; van den Ouweland, A.M.; Beverloo, H.B.; Arentsen-Peters, S.T.; Zwaan, C.M.; Wagner, A. Acute myeloid leukaemia in a case with Tatton-Brown-Rahman syndrome: The peculiar DNMT3A R882 mutation. *J. Med. Genet.* **2017**, *54*, 805–808. [CrossRef]
69. Kosaki, R.; Terashima, H.; Kubota, M.; Kosaki, K. Acute myeloid leukemia-associated DNMT3A p. Arg882His mutation in a patient with Tatton-Brown–Rahman overgrowth syndrome as a constitutional mutation. *Am. J. Med. Genet. Part A* **2017**, *173*, 250–253. [CrossRef]
70. Tlemsani, C.; Luscan, A.; Leulliot, N.; Bieth, E.; Afenjar, A.; Baujat, G.; Doco-Fenzy, M.; Goldenberg, A.; Lacombe, D.; Lambert, L.; et al. SETD2 and DNMT3A screen in the Sotos-like syndrome French cohort. *J. Med. Genet.* **2016**, *53*, 743–751. [CrossRef]
71. Xin, B.; Marino, T.C.; Szekely, J.; Leblanc, J.; Cechner, K.; Sency, V.; Wensel, C.; Barabas, M.; Therriault, V.; Wang, H. Novel *DNMT3A* germline mutations are associated with inherited Tatton-Brown-Rahman syndrome. *Clin. Genet.* **2017**, *91*, 623–628. [CrossRef]
72. Shen, W.; Heeley, J.M.; Carlston, C.M.; Acuna-Hidalgo, R.; Nillesen, W.M.; Dent, K.M.; Douglas, G.V.; Levine, K.L.; Bayrak-Toydemir, P.; Marcelis, C.L.; et al. The spectrum of DNMT3A variants in Tatton-Brown-Rahman syndrome overlaps with that in hematologic malignancies. *Am. J. Med. Genet. Genet. A* **2017**, *173*, 3022–3028. [CrossRef] [PubMed]
73. Jeffries, A.R.; Maroofian, R.; Salter, C.G.; Chioza, B.A.; Cross, H.E.; Patton, M.A.; Temple, I.K.; Mackay, D.; Rezwan, F.I.; Aksglaede, L.; et al. Growth disrupting mutations in epigenetic regulatory molecules are associated with abnormalities of epigenetic aging. *bioRxiv* **2018**, 477356. [CrossRef]
74. Zhang, Z.-M.; Lü, R.; Wang, P.; Yu, Y.; Chen, D.-L.; Gao, L.; Liu, S.; Ji, D.; Rothbart, S.B.; Wang, Y.; et al. Structural basis for DNMT3A-mediated de novo DNA methylation. *Nat. Cell Boil.* **2018**, *554*, 387–391. [CrossRef]
75. Ley, T.J.; Ding, L.; Walter, M.J.; McLellan, M.D.; Lamprecht, T.; Larson, D.E.; Kandoth, C.; Payton, J.E.; Baty, J.; Welch, J.; et al. *DNMT3A* Mutations in Acute Myeloid Leukemia. *New Engl. J. Med.* **2010**, *363*, 2424–2433. [CrossRef] [PubMed]
76. Russler-Germain, D.A.; Spencer, D.H.; Young, M.A.; Lamprecht, T.L.; Miller, C.A.; Fulton, R.; Meyer, M.R.; Erdmann-Gilmore, P.; Townsend, R.R.; Wilson, R.K.; et al. The R882H DNMT3A mutation associated with AML dominantly inhibits wild-type DNMT3A by blocking its ability to form active tetramers. *Cancer Cell* **2014**, *25*, 442–454. [CrossRef] [PubMed]
77. Kim, S.J.; Zhao, H.; Hardikar, S.; Singh, A.K.; Goodell, M.A.; Chen, T. A DNMT3A mutation common in AML exhibits dominant-negative effects in murine ES cells. *Blood* **2013**, *122*, 4086–4089. [CrossRef] [PubMed]
78. Remacha, L.; Currás-Freixes, M.; Torres-Ruiz, R.; Schiavi, F.; Torres-Pérez, R.; Calsina, B.; Letón, R.; Comino-Mendez, I.; Roldan-Romero, J.M.; Montero-Conde, C.; et al. Gain-of-function mutations in DNMT3A in patients with paraganglioma. *Genet. Med.* **2018**, *20*, 1644–1651. [CrossRef]

79. Averbuch, S.D.; Steakley, C.S.; Young, R.C.; Gelmann, E.P.; Goldstein, D.S.; Stull, R.; Keiser, H.R. Malignant pheochromocytoma: Effective treatment with a combination of cyclophosphamide, vincristine, and dacarbazine. *Ann. Intern. Med.* **1988**, *109*, 267–273. [CrossRef] [PubMed]
80. Heyn, P.; Logan, C.V.; Fluteau, A.; Challis, R.C.; Auchynnikava, T.; Martin, C.A.; Marsh, J.A.; Taglini, F.; Kilanowski, F.; Parry, D.A.; et al. Gain-of-function *DNMT3A* mutations cause microcephalic dwarfism and hypermethylation of Polycomb-regulated regions. *Nat. Genet.* **2019**, *51*, 96–105. [CrossRef]
81. Xie, W.; Schultz, M.D.; Lister, R.; Hou, Z.; Rajagopal, N.; Ray, P.; Whitaker, J.W.; Tian, S.; Hawkins, R.D.; Leung, D.; et al. Epigenomic analysis of multi-lineage differentiation of human embryonic stem cells. *Cell* **2013**, *153*, 1134–1148. [CrossRef]
82. Li, Y.; Zheng, H.; Wang, Q.; Zhou, C.; Wei, L.; Liu, X.; Zhang, W.; Zhang, Y.; Du, Z.; Wang, X.; et al. Genome-wide analyses reveal a role of Polycomb in promoting hypomethylation of DNA methylation valleys. *Genome Boil.* **2018**, *19*, 18. [CrossRef]
83. Jeong, M.; Sun, D.; Luo, M.; Huang, Y.; Challen, G.A.; Rodriguez, B.; Zhang, X.; Chavez, L.; Wang, H.; Hannah, R.; et al. Large conserved domains of low DNA methylation maintained by Dnmt3a. *Nat. Genet.* **2014**, *46*, 17. [CrossRef]
84. Long, H.K.; Sims, D.; Heger, A.; Blackledge, N.P.; Kutter, C.; Wright, M.L.; Grützner, F.; Odom, D.T.; Patient, R.; Ponting, C.P.; et al. Epigenetic conservation at gene regulatory elements revealed by non-methylated DNA profiling in seven vertebrates. *eLife* **2013**, *2*, 00348. [CrossRef]
85. Wu, H.; Coskun, V.; Tao, J.; Xie, W.; Ge, W.; Yoshikawa, K.; Li, E.; Zhang, Y.; Sun, Y.E. Dnmt3a-dependent nonpromoter DNA methylation facilitates transcription of neurogenic genes. *Science* **2010**, *329*, 444–448. [CrossRef]
86. Challen, G.A.; Sun, D.; Jeong, M.; Luo, M.; Jelinek, J.; Berg, J.S.; Bock, C.; VasanthaKumar, A.; Gu, H.; Xi, Y.; et al. Dnmt3a is essential for hematopoietic stem cell differentiation. *Nat. Genet.* **2011**, *44*, 23–31. [CrossRef]
87. Shirohzu, H.; Kubota, T.; Kumazawa, A.; Sado, T.; Chijiwa, T.; Inagaki, K.; Suetake, I.; Tajima, S.; Wakui, K.; Miki, Y.; et al. Three novelDNMT3B mutations in Japanese patients with ICF syndrome. *Am. J. Med. Genet.* **2002**, *112*, 31–37. [CrossRef] [PubMed]
88. Ge, Y.-Z.; Gowher, H.; Jeltsch, A.; Pu, M.-T.; Wu, H.-P.; Ding, J.-P.; Xu, G.-L. Chromatin targeting of de novo DNA methyltransferases by the PWWP domain. *J. Boil. Chem.* **2004**, *279*, 25447–25454. [CrossRef]
89. Baubec, T.; Colombo, D.F.; Wirbelauer, C.; Schmidt, J.; Burger, L.; Krebs, A.R.; Akalin, A.; Schübeler, D. Genomic profiling of DNA methyltransferases reveals a role for DNMT3B in genic methylation. *Nat. Cell Boil.* **2015**, *520*, 243–247. [CrossRef]
90. Duymich, C.E.; Charlet, J.; Yang, X.; Jones, P.A.; Liang, G. DNMT3B isoforms without catalytic activity stimulate gene body methylation as accessory proteins in somatic cells. *Nat. Commun.* **2016**, *7*, 11453. [CrossRef]
91. Van Emburgh, B.O.; Robertson, K.D. Modulation of Dnmt3b function in vitro by interactions with Dnmt3L, Dnmt3a and Dnmt3b splice variants. *Nucleic Acids* **2011**, *39*, 4984–5002. [CrossRef]
92. Jin, B.; Tao, Q.; Peng, J.; Soo, H.M.; Wu, W.; Ying, J.; Fields, C.R.; Delmas, A.L.; Liu, X.; Qiu, J.; et al. DNA methyltransferase 3B (DNMT3B) mutations in ICF syndrome lead to altered epigenetic modifications and aberrant expression of genes regulating development, neurogenesis and immune function. *Hum. Mol. Genet.* **2008**, *17*, 690–709. [CrossRef]
93. Maresca, A.; Zaffagnini, M.; Caporali, L.; Carelli, V.; Zanna, C. DNA methyltransferase 1 mutations and mitochondrial pathology: Is mtDNA methylated? *Front. Genet.* **2015**, *6*, 90. [CrossRef] [PubMed]
94. Schlesinger, Y.; Straussman, R.; Keshet, I.; Farkash, S.; Hecht, M.; Zimmerman, J.; Eden, E.; Yakhini, Z.; Ben-Shushan, E.; Reubinoff, B.E.; et al. Polycomb-mediated methylation on Lys27 of histone H3 pre-marks genes for de novo methylation in cancer. *Nat. Genet.* **2007**, *39*, 232–236. [CrossRef] [PubMed]

© 2019 by the authors. Licensee MDPI, Basel, Switzerland. This article is an open access article distributed under the terms and conditions of the Creative Commons Attribution (CC BY) license (http://creativecommons.org/licenses/by/4.0/).

MDPI
St. Alban-Anlage 66
4052 Basel
Switzerland
Tel. +41 61 683 77 34
Fax +41 61 302 89 18
www.mdpi.com

Genes Editorial Office
E-mail: genes@mdpi.com
www.mdpi.com/journal/genes

www.ingramcontent.com/pod-product-compliance
Lightning Source LLC
LaVergne TN
LVHW071955080526
838202LV00064B/6755